BARRON'S

EARLY ACHIEVER

GRADE 3

MATH WORKBOOK

ACTIVITIES & PRACTICE

REVIEW • UNDERSTAND • DISCOVER

Copyright © 2022 by Kaplan North America, LLC, d/b/a Barron's Educational Series

Previously published under the title *Barron's Common Core Success Grade 3 Math*
by Kaplan North America, LLC, d/b/a Barron's Educational Series

All rights reserved.
No part of this publication may be reproduced or distributed in any form
or by any means without the written permission of the copyright owner.

Published by Kaplan North America, LLC, d/b/a Barron's Educational Series
1515 W Cypress Creek Road
Fort Lauderdale, FL 33309
www.barronseduc.com

ISBN 978-1-5062-8155-1

10 9 8 7 6 5 4 3 2 1

Kaplan North America, LLC, d/b/a Barron's Educational Series print books are available at special quantity discounts to use for sales promotions, employee premiums, or educational purposes. For more information or to purchase books, please call the Simon & Schuster special sales department at 866-506-1949.

Credits: Page 2 ©Matthias G. Ziegler/Shutterstock, Page 6 ©xenia_ok/Shutterstock, Page 27 (all) ©firedark /Shutterstock, Page 48 ©Vextok/Shutterstock, Page 51 ©SlipFloat/Shutterstock (cards), ©78studion/Shutterstock (album page), Page 53 ©SlipFloat/Shutterstock (cards), Page 62 ©Marish/Shutterstock, Page 71 ©BlueRingMedia/Shutterstock (tubs), BlueRingMedia/Shutterstock, Page 73 ©Mikael Damkier/Shutterstock, Page 74 ©abramsdesign/Shutterstock, Page 76 ©ultramanjack/Shutterstock (brownies), ©Ali Ozgon/Shutterstock (nuts), Page 89 ©abramsdesign/Shutterstock, Page 95 Quang Ho/Shutterstock, Page 102 © greiss desian/Shutterstock, Page 107 ©jeedlove/Shutterstock, 108 ©jeedlove/Shutterstock, Page 109 ©jeedlove/Shutterstock, Page 112 ©Alhovik/Shutterstock (ruler), ©Viktorija Reuta/Shutterstock (crayon), ©Nikiteev_Konstantin/Shutterstock (water bottle), ©chpavel /Shutterstock (coffee cup), Savchenko Liudmyla /Shutterstock (bowl of soup), Page 113 ©La Gorda/Shutterstock (fish), ©Geanine87/Shutterstock (orange), Page 114 (all) ©jeedlove/Shutterstock, Page 116 ©Alhovik/Shutterstock (ruler), ©illustratiostock/Shutterstock(pen), ©La Gorda/Shutterstock (toolbox), ©Vector.design/Shutterstock(tube), ©maglyvi/Shutterstock(comb), ©Geanine87/Shutterstock(apple), ©Lilu330/Shutterstock (dog), ©iunewind/Shutterstock(cell phone), ©Olga Maslov/Shutterstock(weights), ©tele52/Shutterstock(iron), ©Tashal/Shutterstock(toothbrush), ©Jane Kelly/Shutterstock(dishsoap), ©Matthew Cole/Shutterstock(sink), ©solar22/Shutterstock (cup water), Page 144 ©kontur-vid/Shutterstock

BARRON'S

Introduction

Barron's Early Achiever workbooks are based on sound educational practices and include both parent-friendly and teacher-friendly explanations of specific learning goals and how students can achieve them through fun and interesting activities and practice. This exciting series mirrors the way mathematics is taught in the classroom. *Early Achiever Grade 3 Math* presents these skills through different units of related materials that reinforce each learning goal in a meaningful way. The Review, Understand, and Discover sections assist parents, teachers, and tutors in helping students apply skills at a higher level. Additionally, students will become familiar and comfortable with the manner of presentation and learning, as this is what they experience every day in the classroom. These factors will help early achievers master the skills and learning goals in math and will also provide an opportunity for parents to play a larger role in their children's education.

Introduction to Problem Solving

For parents, tutors, teachers, and homework helpers: This book will help to equip both students and parents with strategies to solve math problems successfully. Problem solving in the mathematics classroom involves more than calculations alone. It involves students' ability to consistently show their reasoning and comprehension skills to model and explain what they have been taught. These skills will form the basis for future success in meeting life's goals. Working through these skills each year through the twelfth grade sets the necessary foundation for collegiate and career success. Students will be better prepared to handle the challenges that await them as they gradually enter into the global marketplace.

Making Sense of the Problem-Solving Process

For students: It is important that you be able to make sense of word problems, write word problems with numbers and symbols, and be able to prove when you are right as well as to know when a mistake happened. You may solve a problem by drawing a model or by using a chart, list, or other tool. When you get your correct answer, you must be able to explain how and why you chose to solve it that way. Every word problem in this workbook allows you to practice these skills, helping to prepare you for the demands of problem solving in your third grade classroom. The first unit of this book discusses the **Ace It Time!** section of each lesson. **Ace It Time!** will help you master these skills.

While Doing Mathematics You Will...

1. Make sense of problems and become a champion in solving them

- Solve problems and discuss how you solved them
- Look for a starting point and plan to solve the problem
- Make sense (meaning) of a problem and search for solutions
- Use concrete objects or pictures to solve problems
- Check over work by asking, "Does this make sense?"
- Plan out a problem-solving approach

2. Reason on concepts and understand that they are measurable

- Understand a number represents a specific quantity
- Connect quantities to written symbols
- Take a word problem and represent it with numbers and symbols
- Know and use different properties of operations
- Connect addition and subtraction to length

3. Construct productive arguments and compare the reasoning of others

- Construct arguments using concrete objects, pictures, drawings, and actions
- Practice having conversations/discussions about math
- Explain your own thinking to others and respond to the thinking of others
- Ask questions to clarify the thinking of others (How did you get that answer? Why is that true?)
- Justify your answer and determine if the thinking of others is correct

4. Model with mathematics

- Determine ways to represent the problem mathematically
- Represent story problems in different ways; examples may include using numbers or words, drawing pictures, using objects, acting out, making a chart or list, writing equations
- Make connections between different representations and explain
- Evaluate your answers and think about whether or not they make sense

5. Use appropriate tools strategically

- Consider available tools when solving math problems
- Choose tools appropriately
- Determine when certain tools might be helpful
- Use technology to help with understanding

6. Attend to detail

- Develop math communication skills by using clear and exact language in your math conversations
- Understand meanings of symbols and label appropriately
- Calculate accurately

7. Look for and make use of structure

- Apply general math rules to specific situations
- Look for patterns or structure to help solve problems
- Adopt mental math strategies based on patterns, such as making ten, fact families, and doubles

8. Look for and express regularity in repeated reasoning

- Notice repeated calculations and look for shortcut methods to solve problems (for example, rounding up and adjusting the answer to compensate for the rounding)
- Evaluate your own work by asking, "Does this make sense?"

Contents

Contents

Mathematical Foundations for Grade 3

Problem-Solving Concepts

FOLLOWING THE OBJECTIVE
You will make sense of word problems and use strategies to solve them.

LEARN IT: You are about to learn a lot of different math strategies. This book will help you master multiplication, division, and fractions. Before starting, let's review some math tricks that work for all types of problems. You will use these tricks in the ***Ace It Time!*** section within each lesson.

odd, even, multiply, skip-count,
equals...

STEP 1: UNDERSTAND

What's the Question?

Math problems can have many steps. Each of the steps is shown on the checklist.

The first step is to read the problem and ask yourself, "What question do I have to answer?" and "Will it take more than one step to solve the problem?"

When you find the question, underline it. Then check "Yes" on the checklist.

ACE IT TIME!

	yes	no
Did you underline the question in the word problem?		
Did you circle the numbers or number words?		
Did you box the clue words that tell you what operation to use?		
Did you use a picture to show your thinking?		
Did you label your numbers and your picture?		
Did you explain your thinking and use math vocabulary words in your explanation?		

PRACTICE: Underline the question.

Example: Todd has three toy cars in his toy chest. Each of the cars has 4 wheels. How many wheels do his cars have altogether?

Will it take more than one step to solve the problem? No.

STEP 2: IDENTIFY

What Numbers or Words Are Needed?

It is very important to locate the numbers you will use to solve the problem. When you find the numbers, circle them. Then check "Yes" on the checklist. *Hint:* Some problems might say "4" while others say "four."

ACE IT TIME!

Checklist	yes	no
Did you underline the question in the word problem?		
Did you circle the numbers or number words?		
Did you box the clue words that tell you what operation to use?		
Did you use a picture to show your thinking?		
Did you label your numbers and your picture?		
Did you explain your thinking and use math vocabulary words in your explanation?		

PRACTICE: Circle the numbers you need to solve the problem.

Example: Todd has three toy cars in his toy chest. Each of the cars has 4 wheels. How many wheels do his cars have altogether?

STEP 3: RECOGNIZE THE SUPPORTING DETAILS

Name the Operation

In every problem, there will be clues that help you figure out if you are adding, subtracting, multiplying, or dividing. Put a box around the clues. Then check "Yes" on the checklist.

ACE IT TIME!

	yes	no
Did you underline the question in the word problem?		
Did you circle the numbers or number words?		
Did you box the clue words that tell you what operation to use?		
Did you use a picture to show your thinking?		
Did you label your numbers and your picture?		
Did you explain your thinking and use math vocabulary words in your explanation?		

PRACTICE: Put a box around the clues.

Example: Todd has three toy cars in his toy chest. Each of the cars has 4 wheels. How many wheels do his cars have altogether?

When you have many groups of equal size, you use multiplication. The words "each," "how many," and "altogether" are clues.

STEPS 4–5: SOLVE AND LABEL

It is important that you connect words in your problem to pictures and numbers. Before solving, you should draw a picture or write a math equation to solve the problem. Make sure to label your pictures and equations. Then check "Yes" on the checklist.

PRACTICE: Draw cars.

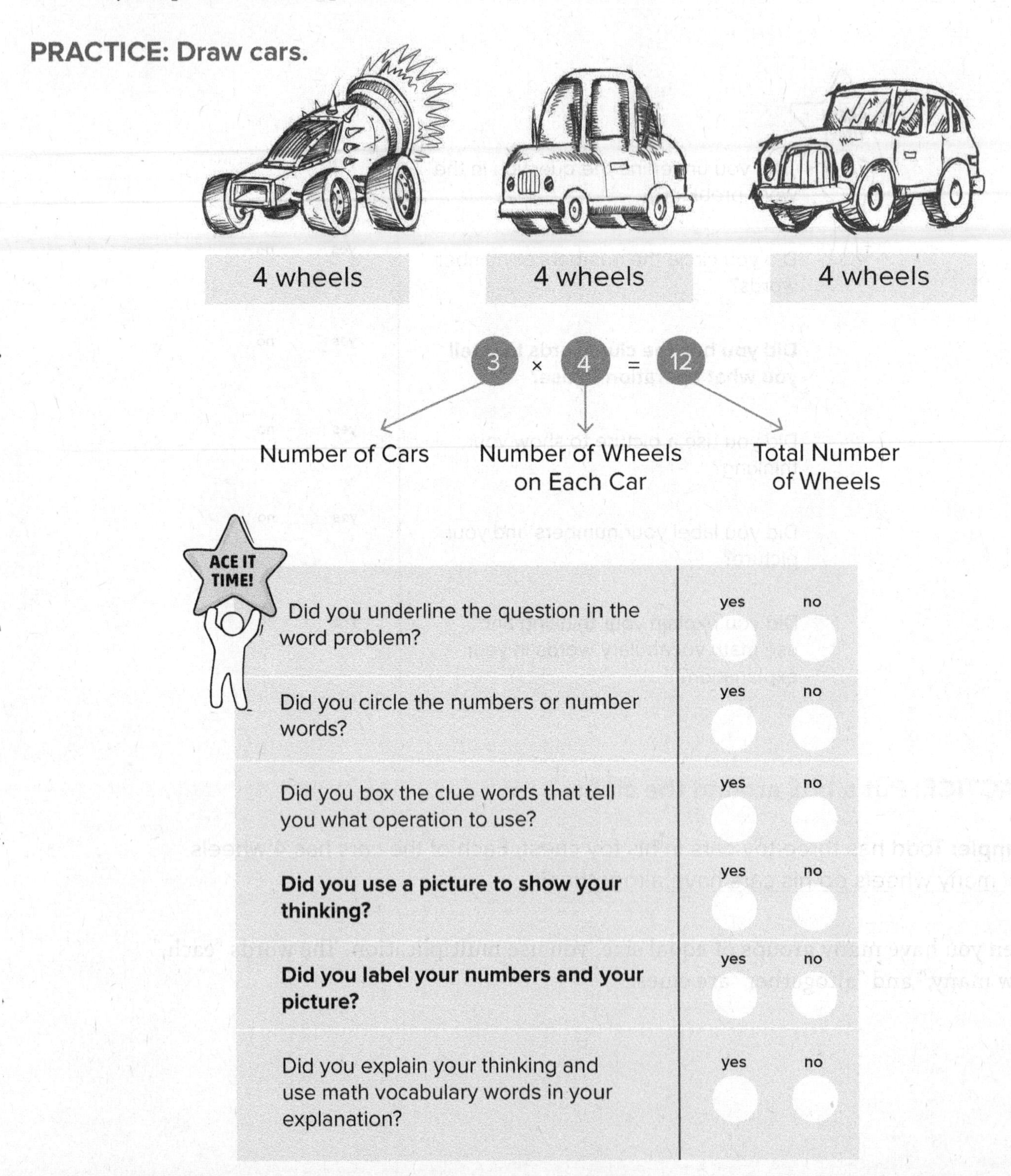

STEP 6: EXPLAIN

Write a Response. Use Math Vocabulary.

You are almost done! Explain your answer and show your thinking. Write in complete sentences to explain the steps you used to solve the problem. Use the vocabulary words in the Math Vocabulary box to help you!

PRACTICE: Explain your answer.

Example: Todd has three toy cars in his toy chest. Each of the cars has 4 wheels. How many wheels do his cars have altogether?

Explanation: First I drew three cars and put 4 wheels on each car. I knew that three times four equals 12 ($3 \times 4 = 12$). I checked my answer by skip-counting by 4s: 4, 8, 12. My answer was 12.

ACE IT TIME!

	yes	no
Did you underline the question in the word problem?		
Did you circle the numbers or number words?		
Did you box the clue words that tell you what operation to use?		
Did you use a picture to show your thinking?		
Did you label your numbers and your picture?		
Did you explain your thinking and use math vocabulary words in your explanation?		

Math Vocabulary

multiply

skip-count

equals

Addition and Subtraction Concepts

Patterns and Properties

FOLLOWING THE OBJECTIVE
You will use patterns to explain properties of addition.

LEARN IT: Look at the addition table. The shaded numbers are ***addends.*** Addends are numbers you add together. The other numbers are ***sums***. Sums are the numbers you get after adding. The table can help you identify ***patterns.*** Patterns are things that repeat and follow rules. Patterns always happen when you add.

Patterns with Zeroes

Look at the examples shaded in dark gray. These sums are made by adding zero to a number.

Use a colored pencil or crayon to color all sums that have 0 as an addend. *Hint:* Start by shading the numbers in row 0.

Do you notice a pattern? What is the rule?

Adding zero shows the **Addition Identity Property**. When you add zero to a number, that number always stays the same.

+	0	1	2	3	4	5	6	7	8	9	10
0	0	1	2	3	4	5	6	7	8	9	10
1	1	2	3	4	5	6	7	8	9	10	11
2	2	3	4	5	6	7	8	9	10	11	12
3	3	4	5	6	7	8	9	10	11	12	13
4	4	5	6	7	8	9	10	11	12	13	14
5	5	6	7	8	9	10	11	12	13	14	15
6	6	7	8	9	10	11	12	13	14	15	16
7	7	8	9	10	11	12	13	14	15	16	17
8	8	9	10	11	12	13	14	15	16	17	18
9	9	10	11	12	13	14	15	16	17	18	19
10	10	11	12	13	14	15	16	17	18	19	20

think!
What happens when you add zero to another number? Does the number change?

Patterns with the Same Addends

Look at the dark gray 2s again. One of them is in row 0, column 2 (0 + 2). One of them is in row 2, column 0 (2 + 0). Is the sum the same? What is the rule?

The Commutative Property says you can write the addends in any order and the answer will be the same.
0 + 2 = 2 and 2 + 0 = 2.

+	0	1	2	3	4	5	6	7	8	9	10
0	0	1	2	3	4	5	6	7	8	9	10
1	1	2	3	4	5	6	7	8	9	10	11
2	2	3	4	5	6	7	8	9	10	11	12
3	3	4	5	6	7	8	9	10	11	12	13
4	4	5	6	7	8	9	10	11	12	13	14
5	5	6	7	8	9	10	11	12	13	14	15
6	6	7	8	9	10	11	12	13	14	15	16
7	7	8	9	10	11	12	13	14	15	16	17
8	8	9	10	11	12	13	14	15	16	17	18
9	9	10	11	12	13	14	15	16	17	18	19
10	10	11	12	13	14	15	16	17	18	19	20

Patterns with Evens and Odds

Use a new color and shade all of the sums of 6 and all of the sums of 9. Compare the addends.

+	0	1	2	3	4	5	6	7	8	9	10
0	0	1	2	3	4	5	6	7	8	9	10
1	1	2	3	4	5	6	7	8	9	10	11
2	2	3	4	5	6	7	8	9	10	11	12
3	3	4	5	6	7	8	9	10	11	12	13
4	4	5	6	7	8	9	10	11	12	13	14
5	5	6	7	8	9	10	11	12	13	14	15
6	6	7	8	9	10	11	12	13	14	15	16
7	7	8	9	10	11	12	13	14	15	16	17
8	8	9	10	11	12	13	14	15	16	17	18
9	9	10	11	12	13	14	15	16	17	18	19
10	10	11	12	13	14	15	16	17	18	19	20

5 (odd) + 1 (odd) = 6 (even)
4 (even) + 2 (even) = 6 (even)
3 (odd) + 3 (odd) = 6 (even)

---

8 (even) + 1 (odd) = 9 (odd)
7 (odd) + 2 (even) = 9 (odd)
6 (even) + 3 (odd) = 9 (odd)

What is the rule? When two even numbers or two odd numbers are added, the sum is even.
(two of the same = even sum)

When an odd number and an even number are added, the sum is odd.
(two different = odd sum)

think! The number 6 is even. Are all of its addends even? Are all of its addends odd? What patterns do you see?

PRACTICE: Now you try

Rewrite each problem using the Commutative Property of Addition.

Sample: $4 + 0 = 0 + 4$ Sum: 4	**1.** $3 + 6 =$ ______ + ______ Sum: ______	**2.** $0 + 8 =$ ______ + ______ Sum: ______

Predict whether the sum will be even or odd.

3. $2 + 2 =$ Even or odd? **4.** $7 + 5 =$ Even or odd? **5.** $2 + 7 =$ Even or odd?

Allison is organizing books on her book shelf. She says she has an odd number of books because she put 9 books on one shelf and 9 books on the other. Do you agree or disagree with Allison? Show your work and explain your thinking on a piece of paper.

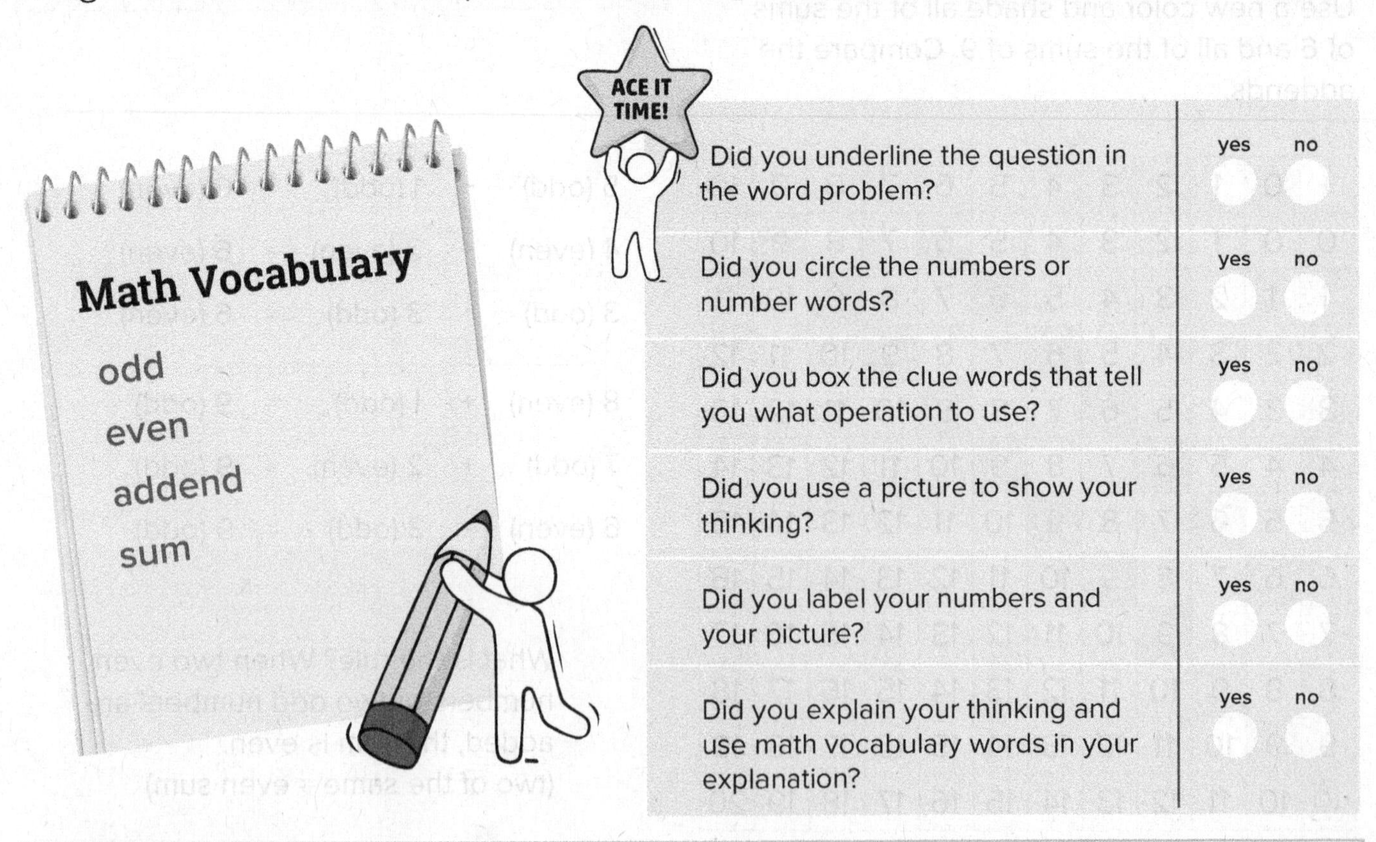

Math on the Move

Play a game of "Odd or Even" to reinforce the concepts learned in this lesson. Using two number cubes or a set of number cards, identify the numbers rolled or drawn as odd or even. Compute the sum and determine if it is odd or even. Record the results on a piece of paper.

Rounding to 10

FOLLOWING THE OBJECTIVE
You will round whole numbers to the nearest 10.

LEARN IT: Counting by ones also increases tens. If you count 1, 2, 3, 4 . . . you will get to 10. If you keep counting, you will get to 20. The number of ones affects the number of tens. For this reason, we use the ones digit to round to the nearest 10.

think! When counting, first you pass 60. Then you count to 70. These are the closest tens.

Example: Round 67 to the nearest 10.

Rounding with Number Lines

Which two tens are closest to 67? Draw them on a number line.

Look at 67. Is it closer to 60 or 70 on the number line? You can use adding to help. You have to add 7 to get from 60 to 67. You only have to add 3 to get from 67 to 70. It is closer to 70.

67 rounds to 70

Wait! What if we were rounding 65? You would add 5 to get from 60 to 65. You would also add the same amount to get from 65 to 70. Which way should you round? Remember: When there are 5 ones, always round up to the higher 10.

think! 5 and above, give it a shove. 4 and below, let it go.

Rounding with a Shortcut

You don't have to draw a number line to round. Let's look at the number 67.

67

Circle the ones place. If the circled digit is less than 5, round down. If the circled digit is 5 or more, round up.

PRACTICE: Now you try

Round each number to the nearest 10 by using the number line.

1. Round 23 = ______________

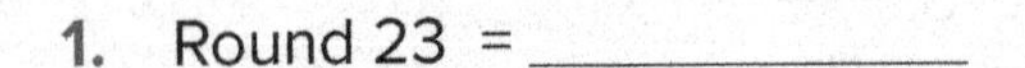

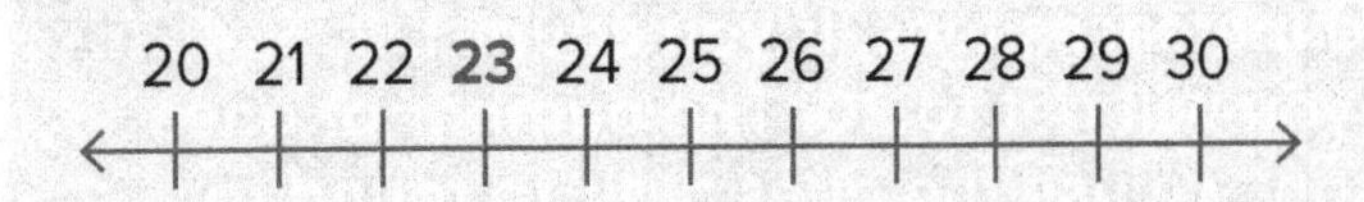

2. Round 48 = ______________

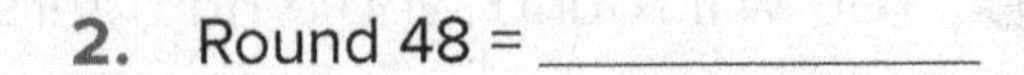

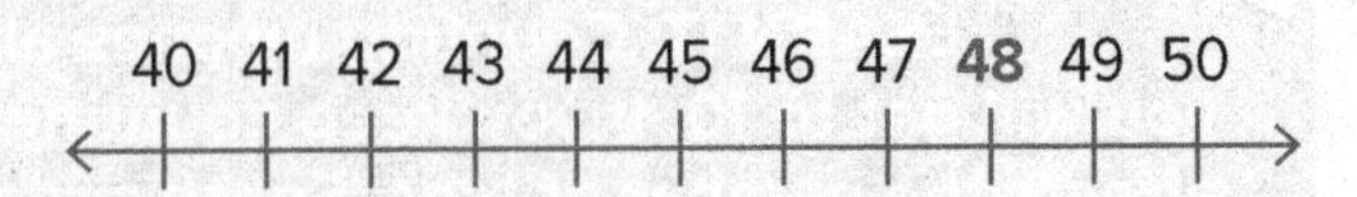

Adam found a math riddle he wanted to solve. It read: "A number when rounded to the nearest ten rounds to 70. Two of these digits (3, 8, 7, or 9) make up the number. What is the number?" Show your work and explain your thinking on a piece of paper.

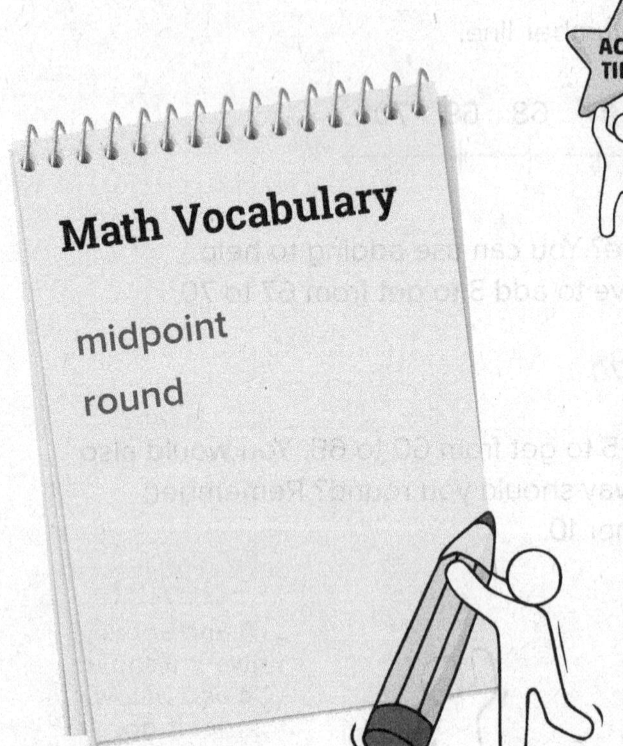

ACE IT TIME!

	yes	no
Did you underline the question in the word problem?		
Did you circle the numbers or number words?		
Did you box the clue words that tell you what operation to use?		
Did you use a picture to show your thinking?		
Did you label your numbers and your picture?		
Did you explain your thinking and use math vocabulary words in your explanation?		

Math on the Move

Rounding can be practiced whenever you read advertisements or see price tags. Round prices up or down. A price of $8.45 rounds up to $8.50. What happens when you see a price of $14.99? Ask an adult or a friend to help you figure it out.

Rounding to 100

FOLLOWING THE OBJECTIVE
You will round whole numbers to the nearest 100.

LEARN IT: Counting by tens increases the hundreds. If you count 10, 20, 30, 40 . . . you will get to 100. If you keep counting, you will get to 200. The number of tens affects the number of hundreds. For this reason, we use the tens digit to round to the nearest 100.

think! When counting, first you pass 200. Then you count to 300. These are the closest hundreds.

Example: Round 279 to the nearest 100.

Rounding with Number Lines

Which two hundreds are closest to 279? Draw them on a number line.

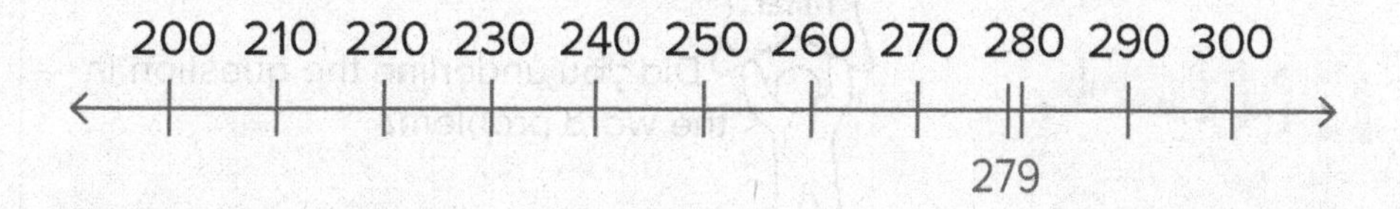

Look at 279. Is it closer to 200 or 300? You can use adding to help. You have to add 79 to get from 200 to 279. You only have to add 21 to get from 279 to 300. It is closer to 300.

279 rounds to 300

Wait! What if we were rounding 250? You would add 50 to get from 200 to 250. You would add the same amount to get from 250 to 300. When there are 5 tens (50), always round up to the higher hundred.

Rounding with a Shortcut

You don't have to draw a number line to round. Let's look at the number 279.

279

Circle the tens place. If the circled digit is less than 5, round down. If the circled digit is 5 or more, round up.

think!
5 and above,
give it a shove.
4 and below,
let it go.

PRACTICE: Now you try

1. Round the number to the nearest 100 by using the number line.

   Round 224 = ______________

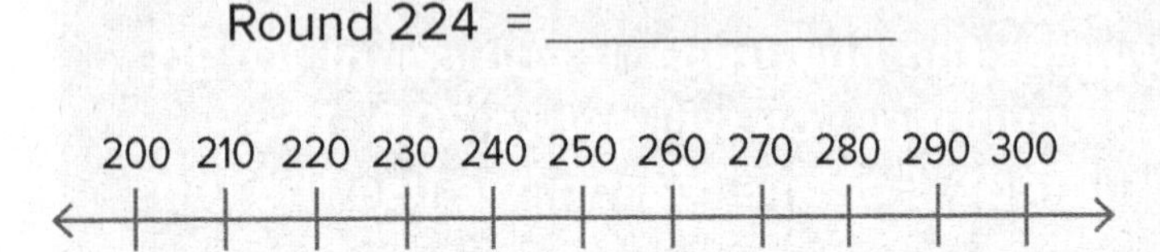

2. Round the number to the nearest 100 using place value.

   279 = ______________

Alex was organizing his baseball card collection. He counted 673 cards. He said he had about 600 cards. His friend Michael disagreed and said he had about 700. Who is correct—Alex or Michael? Show your work and explain your thinking on a piece of paper.

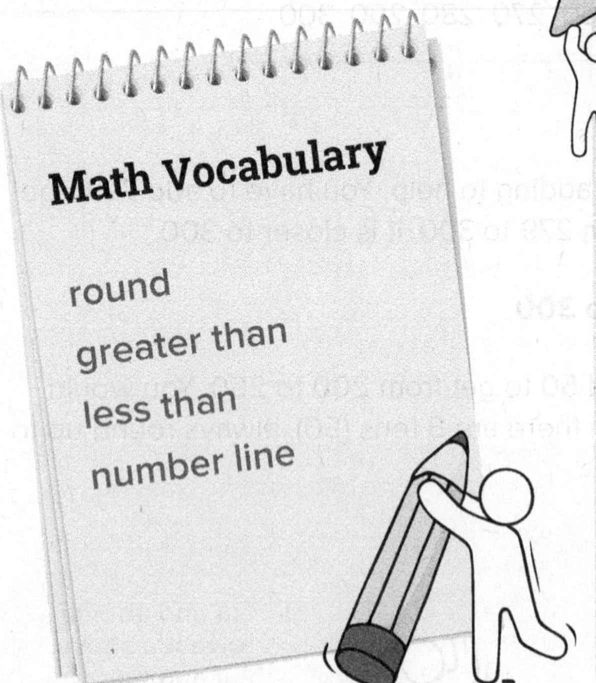

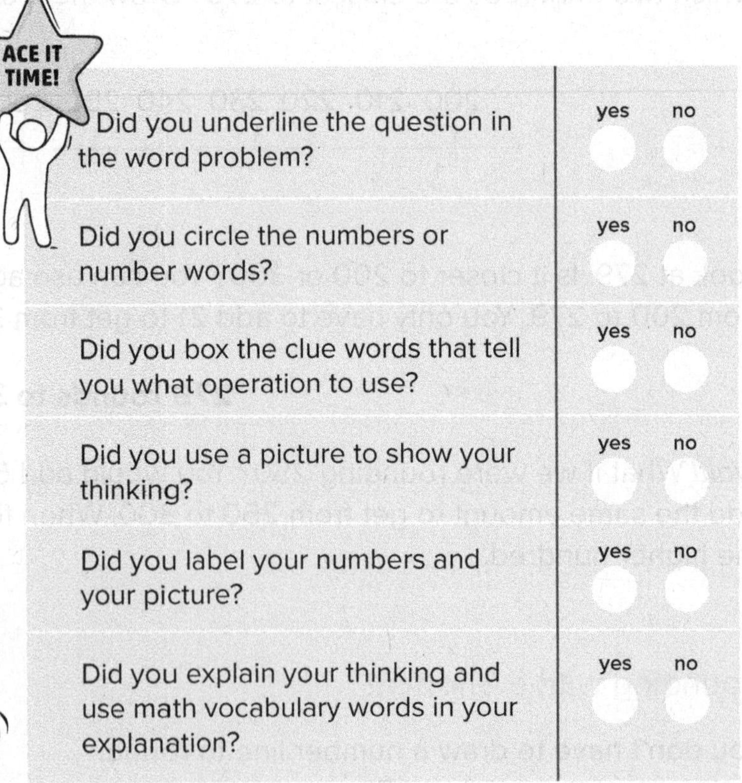

Math on the Move

Ask an adult or a friend to play a game. Get a set of number cards. Draw 3 cards each. Rearrange your cards to make the highest number possible. Round that number to the nearest 100. Whoever makes the highest rounded 100 wins.

Methods of Addition

FOLLOWING THE OBJECTIVE
You will add numbers up to 1,000 using four different methods.

LEARN IT: There are four methods for adding numbers. These help you regroup the ***addends*** (numbers being added) so that addition is easier.

Break Apart Strategy

Break each addend into its place values. The number 466 has 4 hundreds, 6 tens, and 6 ones. These are its parts.

Add the parts together to find **partial sums**. Add the partial sums together to find the total.

466 → 400 + 60 + 6
+315 → 300 + 10 + 5

Partial Sums: 700 70 11

Total Sum: 700 + 70 + 10 + 1 = 781

think! You can break apart partial sums too.

Place Value Strategy

Write the addends vertically. Line up each number by place value. Add the ones, then the tens, and then the hundreds.

Add ones	Add tens	Add hundreds
1 466 +315 1	1 466 +315 81	1 466 +315 781

think! Since 6 + 5 = 11, that means there is 1 ten and 1 one. Why do we carry the ten to the next column?

Use of Properties

Regroup numbers so they are easier to add. *Hint:* 5s and 10s are easiest to add.

think! Which properties are you using?

27
44
16
+ 33

Change the order to make tens

2
27
33 → 10
16
+ 44 → 10
120

Add 10 + 10 = 20
Carry the 2

Use Compatible Numbers

Compatible numbers are numbers that are easy to add in your head. Try numbers like 25, 50, 75, or 100. Break each addend into compatible numbers.

466 → 450 + 10 + 6
+315 → 300 + 10 + 5

Partial Sums: 750 + 20 + 11 = 781

think! How is this method like the Break Apart method? How is it different?

PRACTICE: Now you try

Find each sum. Use a different method for each problem. Show your work on a piece of paper.

1. $\begin{array}{r} 42 \\ 29 \\ 38 \\ +11 \\ \hline \end{array}$ I used the ____________ method.	**2.** $\begin{array}{r} 436 \\ +553 \\ \hline \end{array}$ I used the ____________ method.

Marcus spent 175 minutes reading this week. He plans to read 175 minutes next week and the week after that. If he reads 175 minutes each week for three weeks, how long will Marcus have spent reading? Show your work and explain your thinking on a piece of paper.

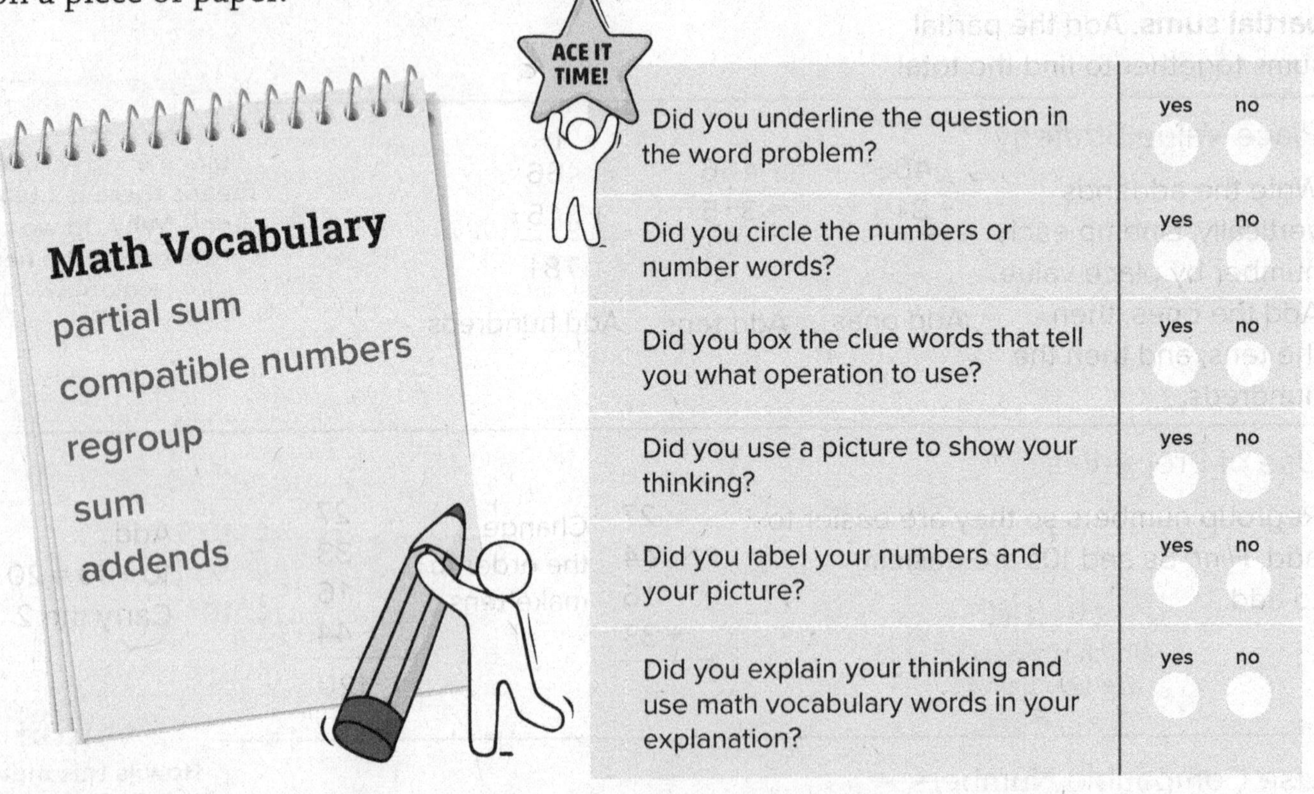

Counting coins is a great way to practice adding. Ask an adult to lend you a random number of coins. Add to find how much money you have. Give the money back when you are done!

Methods of Subtraction

FOLLOWING THE OBJECTIVE
You will subtract numbers up to 1,000 using different methods.

LEARN IT: There are four methods for subtracting numbers. These help you regroup the numbers being subtracted so the math is easier.

Break Apart Strategy

Break numbers into their place value parts. Subtract the parts to find partial differences. Add the partial differences to find the total difference.

$$\begin{array}{rcl} 375 & \rightarrow & 300 + 70 + 5 \\ -111 & \rightarrow & 100 + 10 + 1 \\ \hline & & 200 + 60 + 4 \end{array}$$

$200 + 60 + 4 = 264$

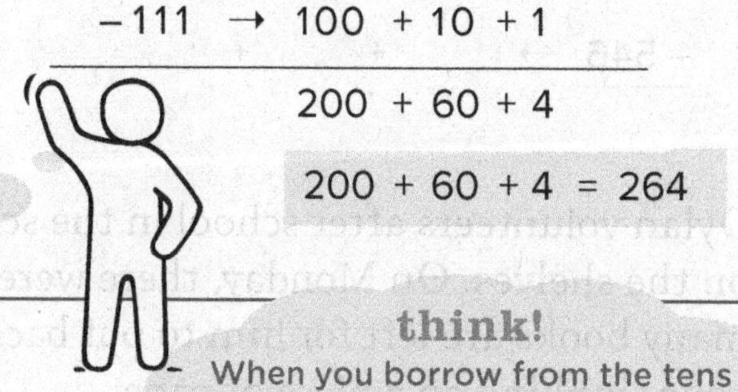

think!
Why do we add partial differences instead of subtracting them? Is the number 375 = 300 - 70 - 5 ?

Place Value Strategy

Write the numbers vertically. Line up each number by place value. Subtract the ones, then the tens, and then the hundreds.

think!
When you borrow from the tens place, you are borrowing a whole ten (10). Why does the number in the tens place only go down by 1 (6 becomes 5)?

Subtract ones	Subtract tens	Subtract hundreds
3 < 5 so regroup	5 < 8 so regroup	4 > 2 so subtract
$\begin{array}{r} 5\ 13 \\ 5\not{6}3 \\ -285 \\ \hline 8 \end{array}$	$\begin{array}{r} 4\ 15\ 13 \\ \not{5}\not{6}3 \\ -285 \\ \hline 78 \end{array}$	$\begin{array}{r} 4\ 15\ 13 \\ \not{5}\not{6}3 \\ -285 \\ \hline 278 \end{array}$

Arrow Method

Begin with the top number and subtract in parts along arrows until you reach the second number. Numbers should be the kind you can easily subtract like 100s, 10s, 5s, and 1s. Add the numbers above the arrows to find the difference.

think!
Since addition and subtraction are opposites of each other, how can you use addition to help you check your work in subtraction?

$$\begin{array}{r} 572 \\ -321 \\ \hline 251 \end{array}$$

$$572 \xrightarrow{-100} 472 \xrightarrow{-100} 372 \xrightarrow{-50} 322 \xrightarrow{-1} 321$$

$$\begin{array}{r} 100 \\ 100 \\ 50 \\ +\ 1 \\ \hline 251 \end{array}$$

Use Compatible Numbers

Remember that compatible numbers are numbers like 25, 50, 75, or 100 that are easy to subtract in your head. Break each number into compatible numbers. Subtract as in the Break Apart Strategy.

655	→	650 + 5	
− 328	→	325 + 3	
		325 + 2	= 327

PRACTICE: Now you try Solve each subtraction problem. Use addition to check.

1. Break Apart Strategy	2. Place Value Strategy	3. Use Compatable Numbers
788 → ____ + ____ + ____ − 546 → ____ + ____ + ____	965 − 138	729 − 526

Dylan volunteers after school in the school library. He helps by putting books back on the shelves. On Monday, there were 153 books to put back. He put back 76. How many books are left for him to put back on the shelves? Show your work and explain your thinking on a piece of paper.

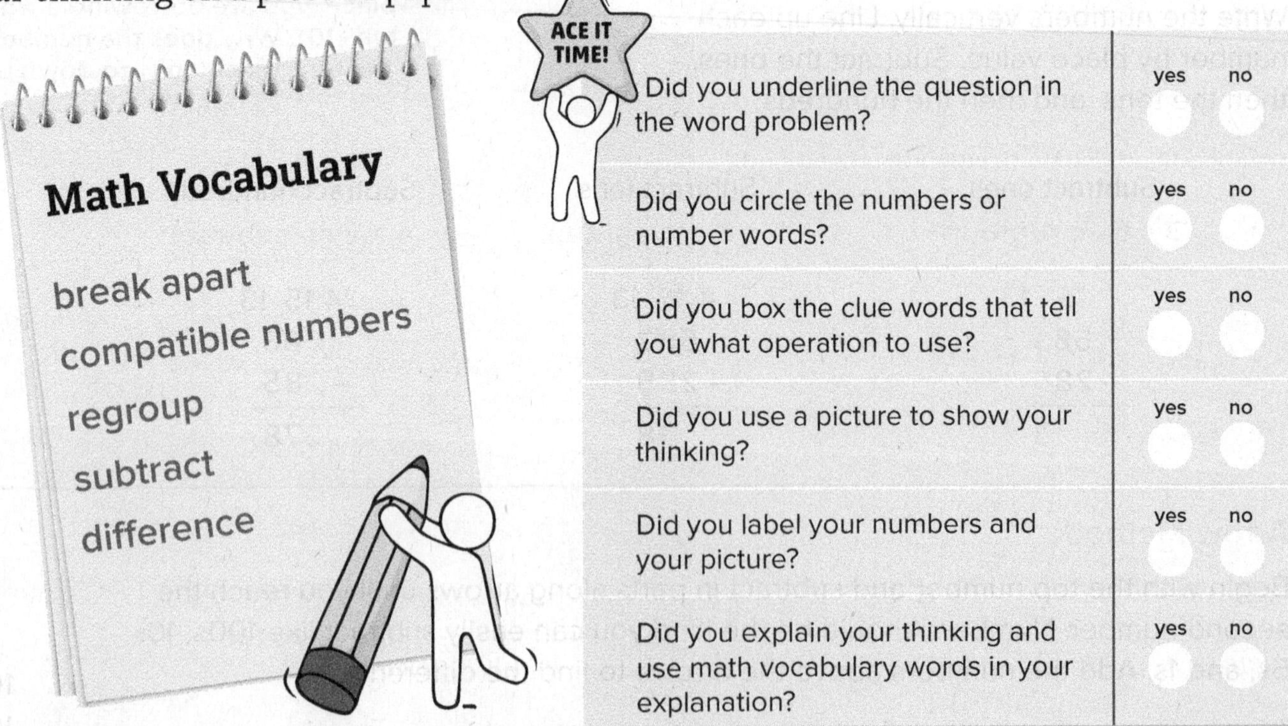

Math on the Move

Ask an adult or a friend to play with you. Draw a line down the middle of a piece of paper to make two columns. Write the number 100 at the top of each column. You use one column and your partner uses the other. Take turns rolling two number cubes and adding the numbers together. Subtract the sum of the cubes from 100. For example, if you roll a 5 and a 6, subtract 11. Play continues until you or your partner gets to 0. The first person to reach 0 wins!

Estimating Sums and Differences

FOLLOWING THE OBJECTIVE
You will use your knowledge of rounding to estimate sums and differences.

LEARN IT: Estimating sums and differences helps you check your work when adding or subtracting.

When Estimating to the Nearest 10	When Estimating to the Nearest 100
Make sure you always round BOTH numbers to the nearest 10.	Make sure you always round BOTH numbers to the nearest 100.
539 → 540 + 372 → + 370 911 910	539 → 500 + 372 → + 400 911 900

PRACTICE: Now you try

Estimate the answer by rounding to **the nearest 10**.

1. 251 → ______ + 328 → + ______	**2.** 338 → ______ − 122 → − ______
3. 273 → ______ + 376 → + ______	**4.** 457 → ______ − 266 → − ______

Estimate the answer by rounding each number to **the nearest 100**.

5.	6.
546 → ________	654 → ________
+178 → + ________	−221 → − ________

A local animal shelter reported there were 176 adoptions in the month of April and 231 adoptions in the month of May. About how many animals were adopted in April and May? Did you round to the nearest 10 or 100? Explain your thinking in rounding to the place value you chose to use on a piece of paper.

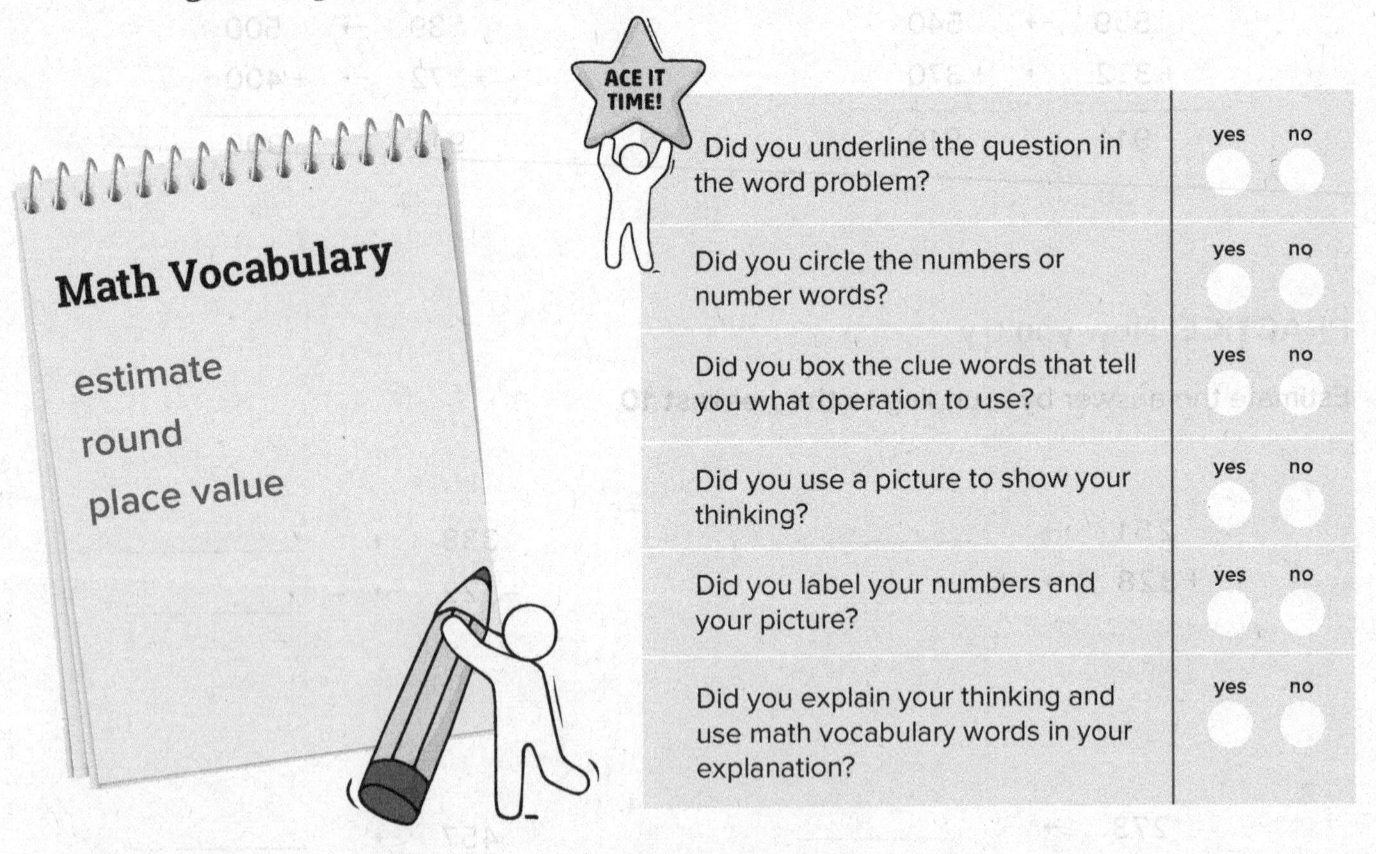

Math on the Move

Think of a number between 0 and 1,000 and round it to the nearest 10. Tell an adult or a friend what the nearest 10 is, and have your partner guess your number. For example, if you say, "My number rounds to 370," your partner could guess 366 or 373. If that is not your number, give more clues, perhaps stating whether your number is greater than or less than your partner's guess. Switch roles when he or she guesses the right number.

REVIEW

Stop and think about what you have learned.

Congratulations! You've finished the lessons for this unit. This means you've noticed patterns in addition. You've learned about properties of addition. You've practiced rounding to tens and hundreds. You can even add and subtract numbers.

Now it's time to prove your addition and subtraction skills. Solve the problems below! Use all of the methods you have learned.

Activity Section 1: Patterns and Properties

Find the sum. Then rewrite the problem using the Commutative Property of Addition.

1. 7 + 0 = ______ ______ + ______ = ______	**2.** 1 + 7 = ______ ______ + ______ = ______	**3.** 0 + 3 = ______ ______ + ______ = ______
4. 10 + 3 = ______ ______ + ______ = ______	**5.** 5 + 0 = ______ ______ + ______ = ______	**6.** 8 + 4 = ______ ______ + ______ = ______

Fill in the blanks to complete the pattern.

7. even + even = ________

8. odd + ________ = even

9. ________ + odd = odd

Activity Section 2: Rounding to 10 and 100

Round each number to the nearest **10**.

1. Round 64 = ________

60 61 62 63 **64** 65 66 67 68 69 70

2. Round 78 = ________

70 71 72 73 74 75 76 77 **78** 79 80

3. Round 46 = ________

40 41 42 43 44 45 **46** 47 48 49 50

4. Round 99 = ________

90 91 92 93 94 95 96 97 98 **99** 100

Round each number to the nearest **100**.

5. Round 624 = ________

600 610 620 630 640 650 660 670 680 690 700

6. Round 397 = ________

300 310 320 330 340 350 360 370 380 390 400

7. Round 166 = ________

100 110 120 130 140 150 160 170 180 190 200

8. Round 711 = ________

700 710 720 730 740 750 760 770 780 790 800

Activity Section 3: Addition and Subtraction Methods

Solve the following addition problems.

1. Break Apart Strategy	2. Place Value Strategy	3. Use Compatible Numbers
274 → _____ + _____ + _____ + 112 → _____ + _____ + _____	5 7 3 + 3 2 8	678 + 126

Solve the following subtraction problems.

4. Break Apart Strategy	5. Place Value Strategy	6. Use Compatible Numbers
274 → _____ + _____ + _____ – 112 → _____ + _____ + _____	8 5 7 – 3 3 8	958 – 427
Use addition to check	Use addition to check	Use addition to check

UNDERSTAND

Understand the meaning of what you have learned and apply your knowledge.

You can use place value understanding to round whole numbers to the nearest 10 or 100.

Activity Section

Mrs. Ortiz challenged her students to discover numbers that will round to 700. She had them use these number cards. They could not repeat a digit in a number, but they could use each card as many times as they needed. She was going to give extra credit to the student who could make 20 or more numbers that would round to 700. Can you meet Mrs. Ortiz's challenge? Show your work to prove each number can round to 700.

0	1	2	3	4
5	6	7	8	9

Numbers that would round to 700 are:

1.	2.	3.	4.	5.
6.	7.	8.	9.	10.
11.	12.	13.	14.	15.
16.	17.	18.	19.	20.

DISCOVER

Discover how you can apply the information you have learned.

You can use place value understanding to round numbers to the nearest 10 or 100. You can also add and subtract up to 1,000.

Activity Section

Mr. Hernandez wants to know how far he drove during his summer vacation to visit family and friends. Answer by completing the table below.

1. Determine the total actual miles driven.
2. Round each distance driven to the nearest 10 and nearest 100.
3. Determine the total miles driven for miles rounded to the nearest 10 and miles rounded to the nearest 100.

Mileage Driven	Nearest 10	Nearest 100
437 miles	____________ miles	____________ miles
289 miles	____________ miles	____________ miles
83 miles	____________ miles	____________ miles
Total ____________ miles	Total ____________ miles	Total ____________ miles

1. Compare the total actual miles driven to both totals of rounded miles driven. Which method—rounding numbers to the nearest 10 or rounding to the nearest 100—gives a total closest to the actual total?

2. Why do you think this is true?

3. Explain how Mr. Hernandez's trip of 83 miles rounds to 100 miles when there is not a digit in the hundreds place.

Multiplication Concepts

Connecting Addition and Multiplication

FOLLOWING THE OBJECTIVE
You will use drawings to show how multiplication and addition are similar.

LEARN IT: Multiplying is a quick way to add groups of the same number.

Example

Mary has a bouquet of six flowers. Each flower has five petals. How many petals are there in all?

Draw a picture to help.

You can count to find the number of petals. You can add 5 + 5 + 5 + 5 + 5 + 5. A faster way is to multiply. Multiplying is a way of adding groups of numbers together. Think, "I will G.E.T. multiplication."

1 — 1 2 3 4 5
2 — 1 2 3 4 5
3 — 1 2 3 4 5
4 — 1 2 3 4 5
5 — 1 2 3 4 5
6 — 1 2 3 4 5

think!
Is it fast to count the petals?
Is it fast to add the number 5 six times?

G.	E.	T.
The first factor is the **group** factor.	The second factor is the **each** factor.	This is the **total**. The total is also called the product.

6 flowers	x	5 petals each	=	30

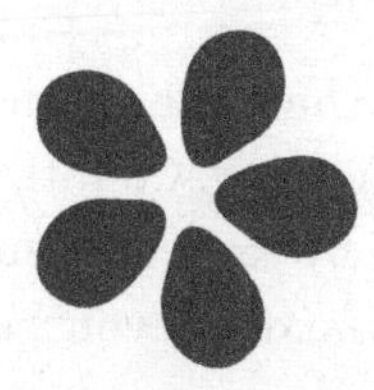

6 groups of 5 = 30
6 × 5 = 30

This is a multiplication fact.
6 × 5 will always equal 30.

PRACTICE: Now you try

Find the answers to the multiplication problems by counting or adding. This will help you learn basic multiplication facts. *Hint:* Each row is a group.

1. 2 groups of 3

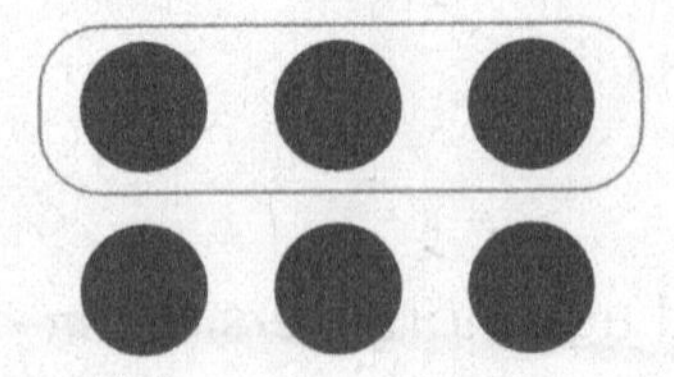

3 + 3

2 × 3 = ___________

2. 3 groups of 5

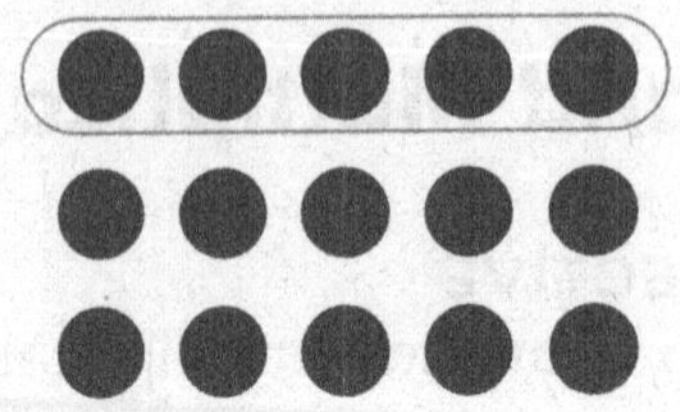

5 + 5 + 5

3 × 5 = ___________

3. 4 groups of 6

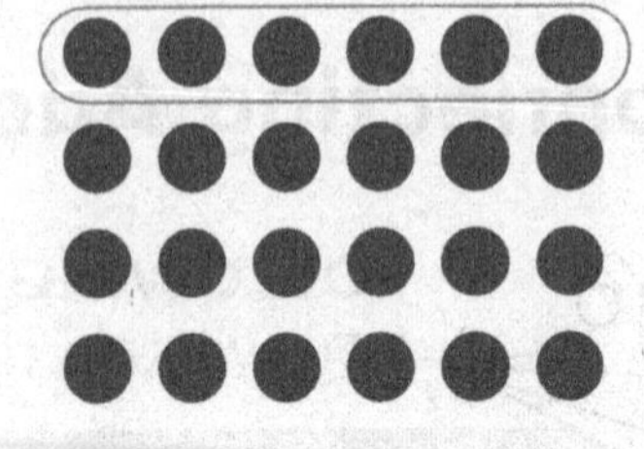

6 + 6 + 6 + 6

4 × 6 = ___________

At the aquarium, Kaitlyn saw 2 horseshoe crabs and 5 octopuses. Horseshoe crabs and octopuses each have 8 legs. How many legs were there in all? Show your work and explain your thinking on a piece of paper.

	yes	no
Did you underline the question in the word problem?		
Did you circle the numbers or number words?		
Did you box the clue words that tell you what operation to use?		
Did you use a picture to show your thinking?		
Did you label your numbers and your picture?		
Did you explain your thinking and use math vocabulary words in your explanation?		

Math Vocabulary

multiply

add

count

Math on the Move

Try counting by tens to 100. This is the same as adding 10 + 10 + 10 and so on to get to 100, which is also the same as multiplying. Counting by numbers will help you learn multiplication facts. After counting by tens, try counting by fives or twos. Then challenge yourself to master the threes, fours, sixes, sevens, eights, and nines.

Skip-Counting

FOLLOWING THE OBJECTIVE
You will use skip-counting to figure out basic multiplication facts.

LEARN IT: Skip-counting means counting by 2s, 3s, 4s, or any number other than 1s. You can show skip-counting on a number line.

Skip-count using a factor of 3, 4, and 5. Fill in the boxes to show counting.

1.

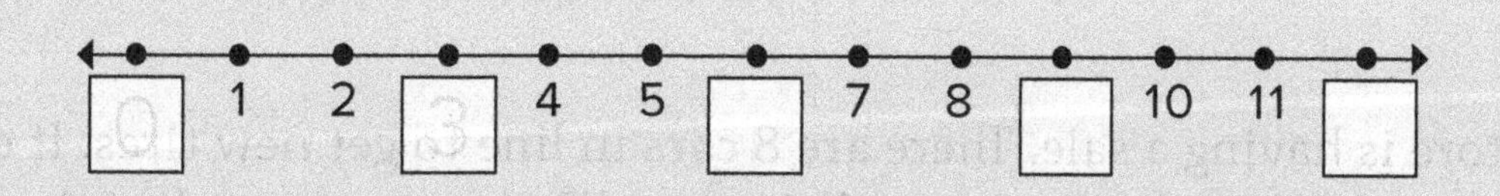

0 . . . 3 . . .

Four skips of 3 is the same as adding four 3s.

This is the same as 4 × 3. **4 × 3** = ____________

2.

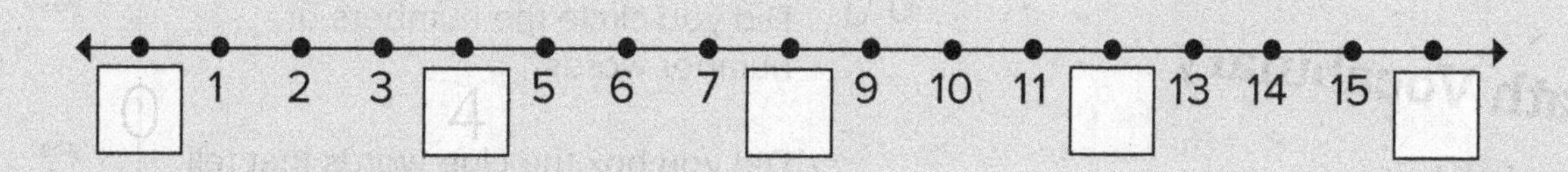

0 . . . 4 . . .

Four skips of 4 is the same as adding four 4s.

This is the same as 4 × 4. **4 × 4** = ____________

3.

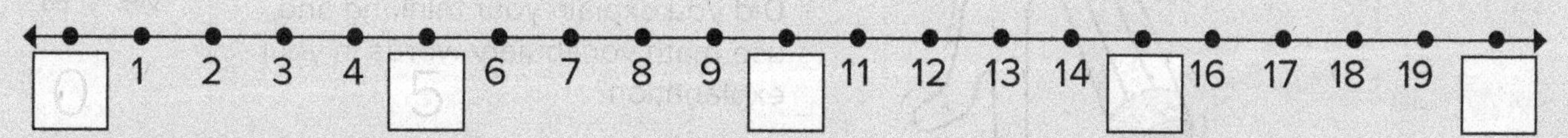

0 . . . 5 . . .

Four skips of 5 is the same as adding four 5s.

This is the same as 4 × 5. **4 × 5** = ____________

PRACTICE: Now you try

Use skip-counting to solve the multiplication problems. The first number is how many times you skip. The second number is what you count by. For example, 2 × 4 means you would make two skips of 4. Remember to always start at zero.

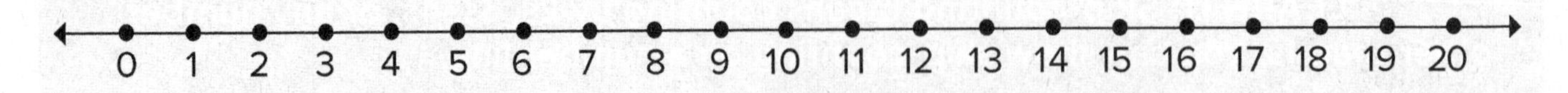

1. 5 × 3 = ____________

   (______ skips of ______)

2. 6 × 3 = ____________

   (______ skips of ______)

Acme Tire Store is having a sale. There are 8 cars in line to get new tires. If each car needs a new set of 4, how many tires will Acme sell? Show your work and explain your thinking on a piece of paper.

ACE IT TIME!

	yes	no
Did you underline the question in the word problem?		
Did you circle the numbers or number words?		
Did you box the clue words that tell you what operation to use?		
Did you use a picture to show your thinking?		
Did you label your numbers and your picture?		
Did you explain your thinking and use math vocabulary words in your explanation?		

Math on the Move

Find a place in your neighborhood, home, or school where you can find numbers to skip-count by. To count the number of eyes you see, count by twos. To count the number of wheels you see on cars, count by fours. What other things can you skip-count?

Multiplying with Arrays

FOLLOWING THE OBJECTIVE
You will use arrays to figure out basic multiplication facts for products less than 100.

LEARN IT: A ***product*** is the number you get when you multiply two numbers. The two numbers you multiply are called ***factors***. ***Arrays*** help you show the factors and figure out the product.

Example: Use an array to solve 4 × 6.

When you see the multiplication symbol (×), think of the words "rows of." The expression 4 × 6 can be read as 4 rows of 6. When you draw the rows, you make an array.

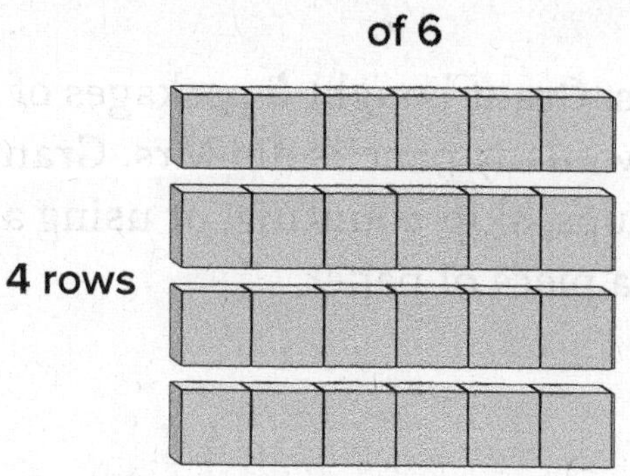

PRACTICE: Now you try

Draw an array to find the product.

1. 4 × 5 = ________	**2.** 3 × 9 = ________	**3.** 5 × 5 = ________
4. 2 × 3 = ________	**5.** 6 × 4 = ________	**6.** 7 × 6 = ________
7. 3 × 4 = ________	**8.** 1 × 10 = ________	**9.** 8 × 4 = ________

Write the multiplication expression shown by each array. Then write the product.

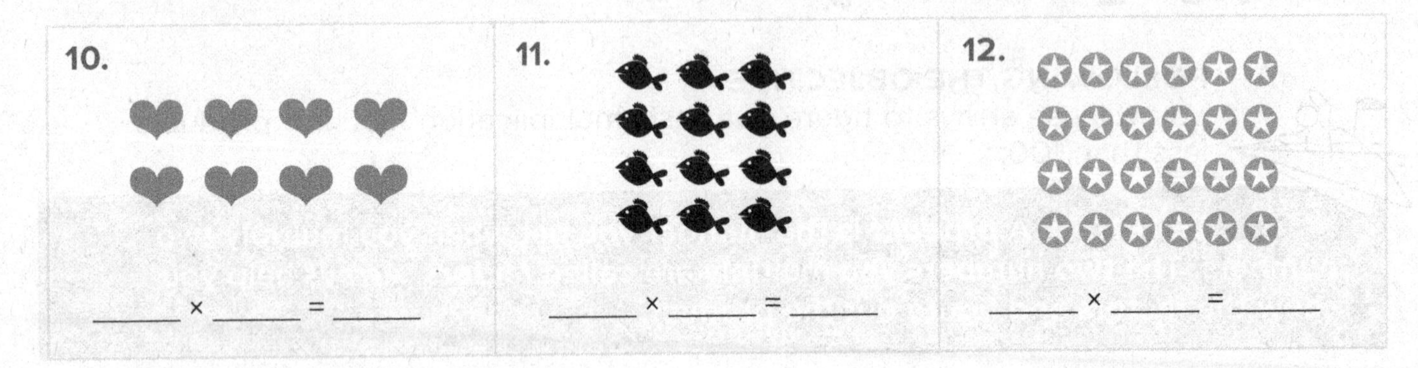

Mrs. Grant bought 6 packages of pencils. There were 8 pencils in each package. How many pencils did Mrs. Grant buy in all? You can solve by adding equal groups, skip-counting, or using arrays. Show your work and explain your thinking on a piece of paper.

Math Vocabulary

- multiplied
- total
- skip-counted
- number line
- array

ACE IT TIME!

	yes	no
Did you underline the question in the word problem?		
Did you circle the numbers or number words?		
Did you box the clue words that tell you what operation to use?		
Did you use a picture to show your thinking?		
Did you label your numbers and your picture?		
Did you explain your thinking and use math vocabulary words in your explanation?		

Math on the Move

Arrays are everywhere! See how many you can find. Postage stamps, sheets of stickers, egg cartons, packages of cupcakes—even floor tiles and bricks are often laid out in arrays. Count the rows and columns and see if you can multiply to find the total number in each array.

The Commutative Property

FOLLOWING THE OBJECTIVE
You will use arrays to show how multiplication and addition have similar properties.

LEARN IT: In addition, it doesn't matter if you write 7 + 3 or 3 + 7. The sum is 10 both ways. This happens because of the Commutative Property of Addition. Multiplication also has a Commutative Property.

Example

Multiply 6 × 2 = 12. This is the same as multiplying 2 × 6 = 12.
Use "rows of" to make an array.
The expression 6 × 2 can be 6 rows of 2.
The expression 2 × 6 can be 2 rows of 6.

Notice how the second array looks like the first array turned sideways. There is the same total!

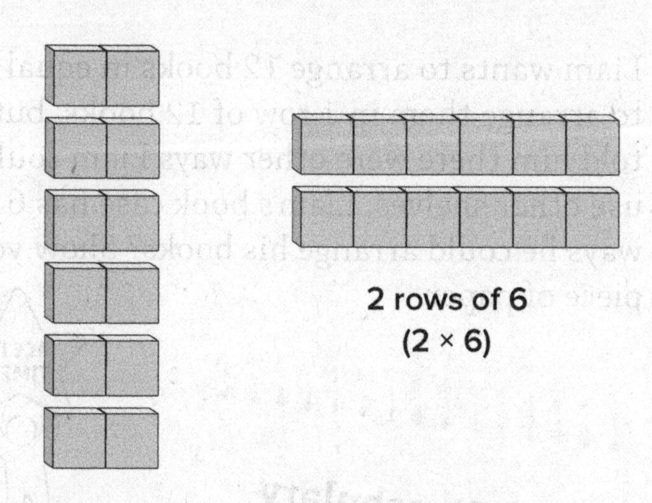

PRACTICE: Now you try

Write a multiplication expression for each array. Use the Commutative Property of Multiplication to write the expression a different way.

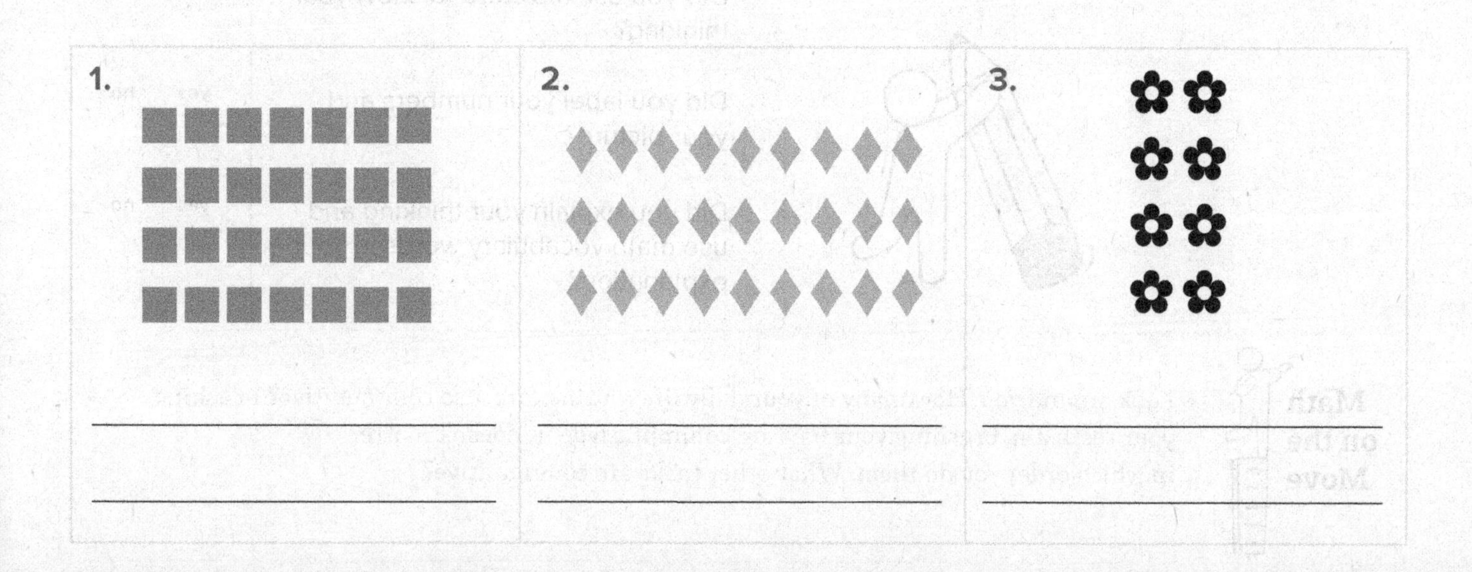

Solve the multiplication problems. Use the Commutative Property of Multiplication to write the problems a different way.

Solve	Related Multiplication Fact
4. 6 × 2 = ________	________
5. 3 × 7 = ________	________
6. 9 × 5 = ________	________

Liam wants to arrange 12 books in equal rows on his book case. First, he tried to arrange them in 1 row of 12 books, but they did not fit on the shelf. His sister told him there were other ways Liam could arrange his books and suggested he use other shelves. Liam's book case has 6 shelves total. What are 4 other possible ways he could arrange his books? Show your work and explain your thinking on a piece of paper.

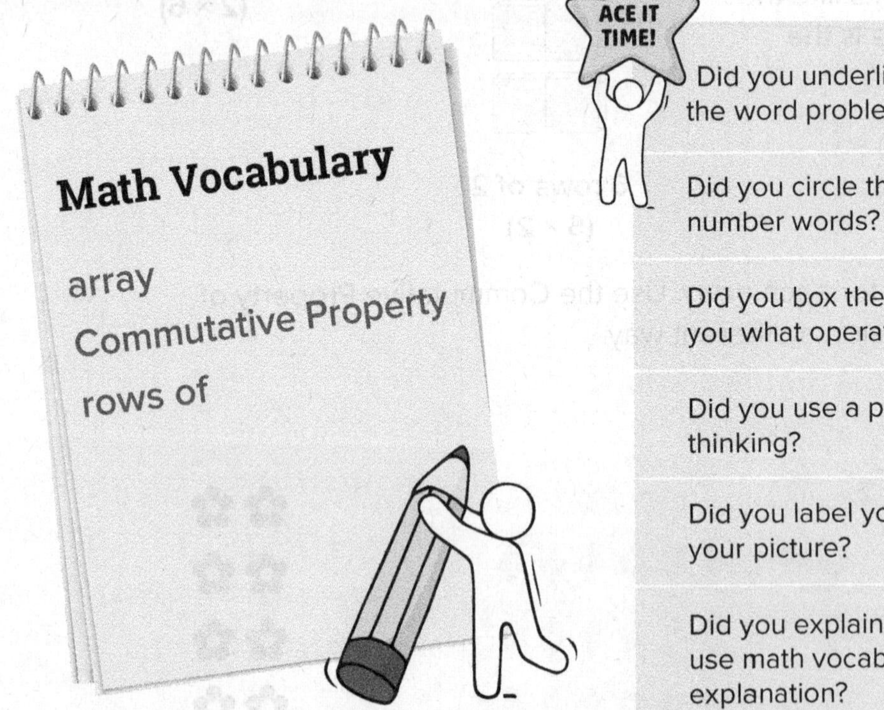

	yes	no
Did you underline the question in the word problem?	○	○
Did you circle the numbers or number words?	○	○
Did you box the clue words that tell you what operation to use?	○	○
Did you use a picture to show your thinking?	○	○
Did you label your numbers and your picture?	○	○
Did you explain your thinking and use math vocabulary words in your explanation?	○	○

Math on the Move

Look around you. How many of your daily life routines are also commutative? Brushing your teeth and brushing your hair are commutative—it doesn't matter in which order you do them. What other tasks are commutative?

Mastering Multiplication

FOLLOWING THE OBJECTIVE
You will multiply numbers up to 100.

LEARN IT: The table below lists the multiplication facts you will need to remember by the end of third grade. There are many things you can do to help you memorize these facts. Practice skip-counting with an adult. Make flash cards. Do whatever works best for you!

0x	1x	2x	3x	4x	5x
0 × 1 = 0	1 × 1 = 1	2 × 1 = 2	3 × 1 = 3	4 × 1 = 4	5 × 1 = 5
0 × 2 = 0	1 × 2 = 2	2 × 2 = 4	3 × 2 = 6	4 × 2 = 8	5 × 2 = 10
0 × 3 = 0	1 × 3 = 3	2 × 3 = 6	3 × 3 = 9	4 × 3 = 12	5 × 3 = 15
0 × 4 = 0	1 × 4 = 4	2 × 4 = 8	3 × 4 = 12	4 × 4 = 16	5 × 4 = 20
0 × 5 = 0	1 × 5 = 5	2 × 5 = 10	3 × 5 = 15	4 × 5 = 20	5 × 5 = 25
0 × 6 = 0	1 × 6 = 6	2 × 6 = 12	3 × 6 = 18	4 × 6 = 24	5 × 6 = 30
0 × 7 = 0	1 × 7 = 7	2 × 7 = 14	3 × 7 = 21	4 × 7 = 28	5 × 7 = 35
0 × 8 = 0	1 × 8 = 8	2 × 8 = 16	3 × 8 = 24	4 × 8 = 32	5 × 8 = 40
0 × 9 = 0	1 × 9 = 9	2 × 9 = 18	3 × 9 = 27	4 × 9 = 36	5 × 9 = 45
0 × 10 = 0	1 × 10 = 10	2 × 10 = 20	3 × 10 = 30	4 × 10 = 40	5 × 10 = 50

6x	7x	8x	9x	10x
6 × 1 = 6	7 × 1 = 7	8 × 1 = 8	9 × 1 = 9	10 × 1 = 10
6 × 2 = 12	7 × 2 = 14	8 × 2 = 16	9 × 2 = 18	10 × 2 = 20
6 × 3 = 18	7 × 3 = 21	8 × 3 = 24	9 × 3 = 27	10 × 3 = 30
6 × 4 = 24	7 × 4 = 28	8 × 4 = 32	9 × 4 = 36	10 × 4 = 40
6 × 5 = 30	7 × 5 = 35	8 × 5 = 40	9 × 5 = 45	10 × 5 = 50
6 × 6 = 36	7 × 6 = 42	8 × 6 = 48	9 × 6 = 54	10 × 6 = 60
6 × 7 = 42	7 × 7 = 49	8 × 7 = 56	9 × 7 = 63	10 × 7 = 70
6 × 8 = 48	7 × 8 = 56	8 × 8 = 64	9 × 8 = 72	10 × 8 = 80
6 × 9 = 54	7 × 9 = 63	8 × 9 = 72	9 × 9 = 81	10 × 9 = 90
6 × 10 = 60	7 × 10 = 70	8 × 10 = 80	9 × 10 = 90	10 × 10 = 100

PRACTICE: Now you try

Solve the multiplication problems.

1. 10 × 6 = ____________ 2. 10 × 9 = ____________ 3. 4 × 7 = ____________

4. 8 × 9 = ____________ 5. 3 × 8 = ____________ 6. 3 × 5 = ____________

7. 5 × 7 = ____________ 8. 2 × 3 = ____________ 9. 3 × 1 = ____________

Yessina and Olivia each have 6 markers. Kenji and Evita each have 9 markers. How many markers do they have in all? Show your work and explain your thinking on a piece of paper.

Math Vocabulary

factor
product
multiply

ACE IT TIME!

	yes	no
Did you underline the question in the word problem?		
Did you circle the numbers or number words?		
Did you box the clue words that tell you what operation to use?		
Did you use a picture to show your thinking?		
Did you label your numbers and your picture?		
Did you explain your thinking and use math vocabulary words in your explanation?		

Math on the Move

Ask an adult or a friend to play a Times Fact War. Use number cards. Deal the entire set face down to yourself and your opponent. Each player turns over one card. Multiply the two cards. The first player to tell the correct product wins the pair. In the event of a tie, each player turns over another card. The fastest one to tell the new product wins both sets of cards. The losing player checks the winner's answer with a calculator. Play continues until one player is out of cards.

The Distributive Property

FOLLOWING THE OBJECTIVE
You will simplify multiplication by breaking factors apart into familiar numbers.

LEARN IT: What happens if you can't remember some of your multiplication tables? You can break apart factors into numbers you know.

think! What other ways can you break up the factor 6? Can you break up the factor 4 instead?

Example: $4 \times 6 = 24$

Select the factor you want to break up. Let's break apart the 6.

What is an addition fact you know that will equal 6?

$4 + 2 = 6$

You can multiply using this addition fact. $4 \times 6 = 4 \times (4 + 2) = (4 \times 4) + (4 \times 2)$. This is called the Distributive Property of Multiplication because you "distribute" the numbers. You can show $4 \times 6 = (4 \times 4) + (4 \times 2)$ using arrays.

Make an array to show 4 rows of 6.

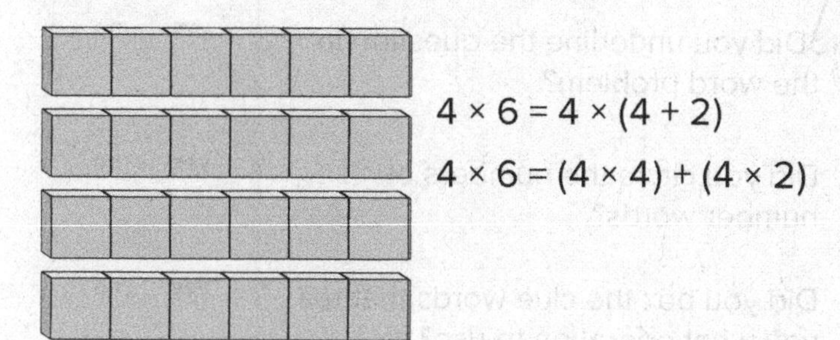

$4 \times 6 = 4 \times (4 + 2)$

$4 \times 6 = (4 \times 4) + (4 \times 2)$

Break apart the array to make two smaller arrays.

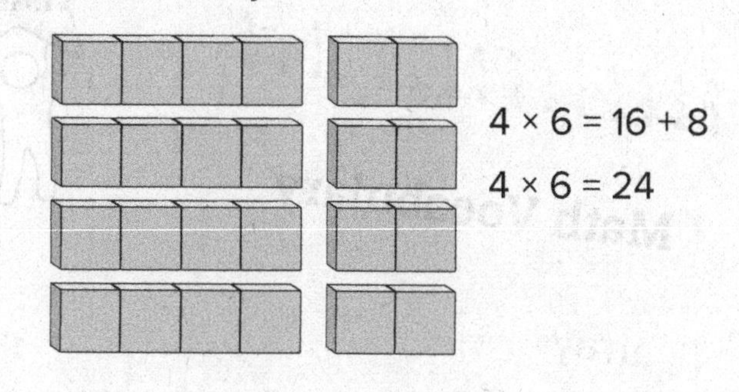

$4 \times 6 = 16 + 8$

$4 \times 6 = 24$

$4 \times 4 =$ **16** $\quad 4 \times 2 =$ **8**

PRACTICE: Now you try

1.

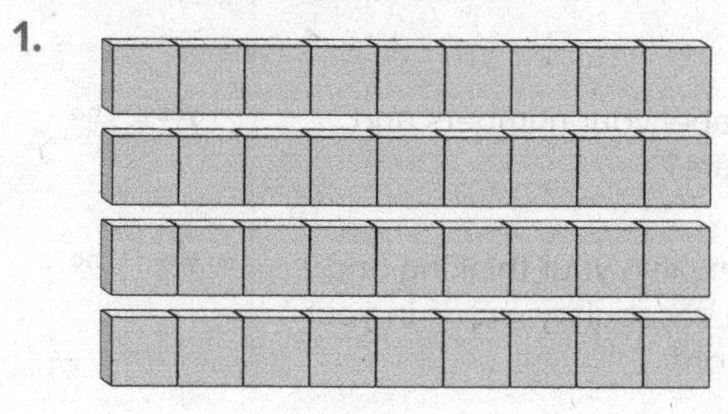

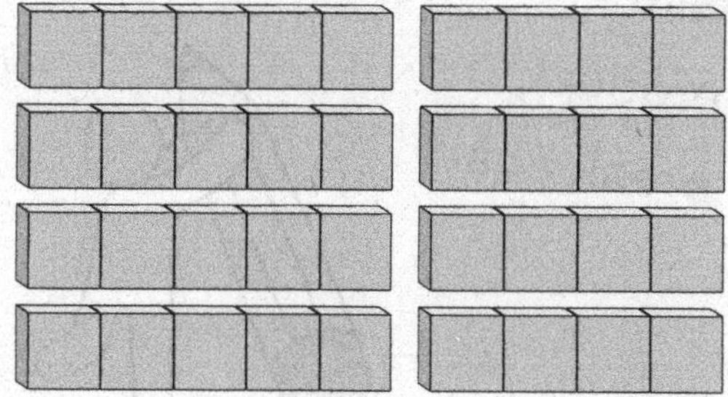

4 rows of 5 + 4 rows of 4

4 rows of 9 is the same as $4 \times 9 =$ ______

$4 \times 9 = 4 \times (5 + 4)$

$4 \times 9 = \quad (4 \times 5) \quad + \quad (4 \times 4)$

$4 \times 9 =$ ________ + ________

$4 \times 9 =$ ________

think! What other ways can you break up the factor 9?

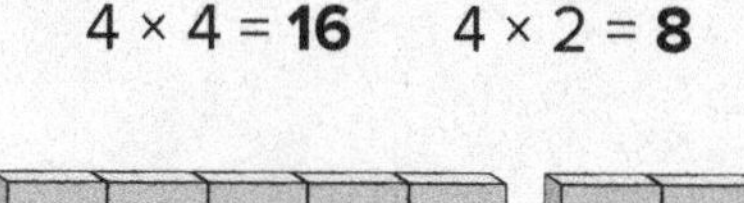

Break the highlighted factor into two different sums. Multiply and solve.

Sample:

8 × 5 = 40	(4 + 4) × 5	(5 + 3) × 5
4 + 4 = 8	(4 × 5) + (4 × 5)	(5 × 5) + (3 × 5)
5 + 3 = 8	20 + 20 = 40	25 + 15 = 40

2. 7 × 6 = _____	(_____ + _____) × 6	(_____ + _____) × 6
_____ + _____ = 7	(_____ × 6) + (_____ × 6)	(_____ × 6) + (_____ × 6)
_____ + _____ = 7	_____ + _____ = _____	_____ + _____ = _____

Martha wants to use the Distributive Property and said that she could solve 9 × 6 by multiplying 4 × 6 and doubling it. Is she correct? Explain. Show your work and explain your thinking on a piece of paper.

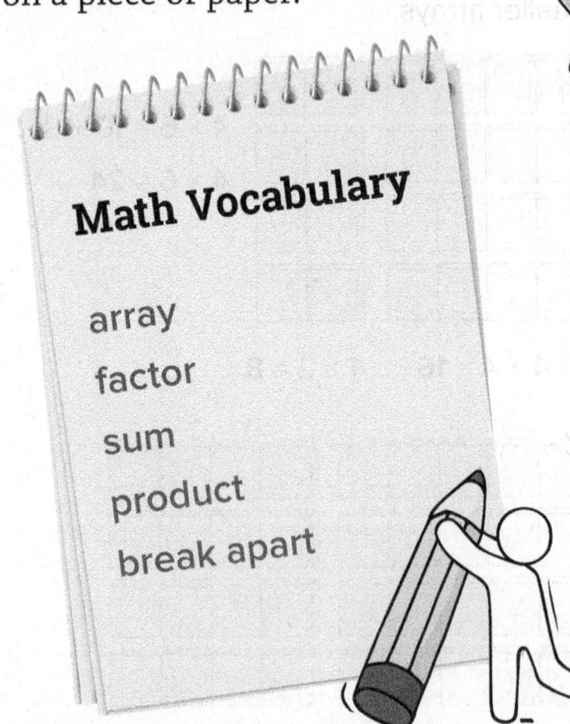

ACE IT TIME!

	yes	no
Did you underline the question in the word problem?	◯	◯
Did you circle the numbers or number words?	◯	◯
Did you box the clue words that tell you what operation to use?	◯	◯
Did you use a picture to show your thinking?	◯	◯
Did you label your numbers and your picture?	◯	◯
Did you explain your thinking and use math vocabulary words in your explanation?	◯	◯

Math on the Move

Learning how to distribute value across coins is a fun way to practice using this property. Have an adult or a friend pick up some change and then say, for example, "I am holding 3 coins that equal 27 cents." What are the coins? Your partner must have a quarter and two pennies. Take turns switching roles.

The Associative Property

FOLLOWING THE OBJECTIVE
You will simplify multiplication by grouping factors together.

LEARN IT: What happens if you have to multiply more than two numbers? You can group them any way you want to make it easier to multiply.

Example: 2 × 2 × 5 = 20

Select two factors you want to multiply first. Ask yourself, "What multiplication fact do I know best?" How about 2 × 2?

Group this fact with parentheses like this: (2 × 2) × 5. Solve the multiplication in parentheses first.

(2 × 2) × 5 = 4 × 5 = 20

PRACTICE: Now you try

Find the product for each. Solve the part in parentheses first and write the new multiplication fact on the first line. Then write the product on the bottom line.

Sample: (4 × 2) × 6 = 4 × (2 × 6) 8 × 6 4 × 12 48 48	**1.** (3 × 3) × 4 = 3 × (3 × 4) ___ × ___ ___ × ___ _____ _____
2. (5 × 2) × 3 = 5 × (2 × 3) ___ × ___ ___ × ___ _____ _____	**3.** (3 × 2) × 2 = 3 × (2 × 2) ___ × ___ ___ × ___ _____ _____
4. (6 × 2) × 5 = 6 × (2 × 5) ___ × ___ ___ × ___ _____ _____	**5.** (4 × 5) × 1 = 4 × (5 × 1) ___ × ___ ___ × ___ _____ _____

Write another way to group the factors. Then find the product.

Sample: (2 × 1) × 7 2 × (1 × 7) 2 × 7 = 14	**6.** 5 × (2 × 5) _____ × _____ _____ × _____ = _____	**7.** 6 × (2 × 5) _____ × _____ _____ × _____ = _____
8. (4 × 1) × 5 _____ × _____ _____ × _____ = _____	**9.** (3 × 2) × 6 _____ × _____ _____ × _____ = _____	**10.** 9 × (2 × 2) _____ × _____ _____ × _____ = _____

Justin was using the Associative Property to solve multiplication sentences with three numbers. His paper got wet, and now he can't read one of the numbers in this math sentence: 7 × (2 × _______) = 56. What is the missing number? Show your work and explain your thinking on a piece of paper.

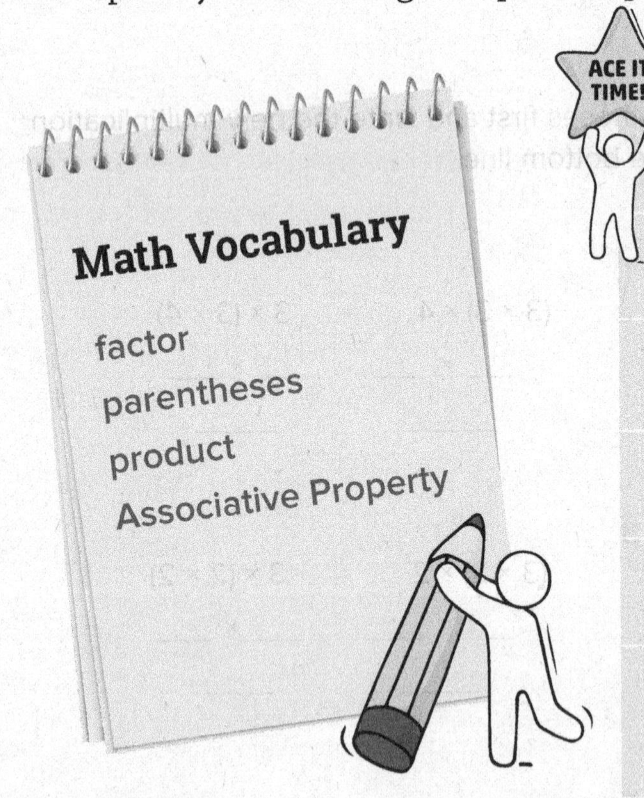

ACE IT TIME!

	yes	no
Did you underline the question in the word problem?		
Did you circle the numbers or number words?		
Did you box the clue words that tell you what operation to use?		
Did you use a picture to show your thinking?		
Did you label your numbers and your picture?		
Did you explain your thinking and use math vocabulary words in your explanation?		

Math on the Move

Roll three number cubes. Multiply the three numbers together. Use the Associative Property to make it easier! Make the game more challenging by playing with four number cubes.

Number Patterns in Multiplication

FOLLOWING THE OBJECTIVE
You will identify and explain patterns in multiplication tables.

LEARN IT: Patterns can help you check your multiplication facts.

How do you find patterns? Compare the numbers being multiplied (factors) with their products. Compare all the products of one factor to each other. *Hint:* The products of the number 2 can all be found in the same column or the same row.

What types of patterns can you find? Maybe all the products of one factor are odd. Maybe the products alternate between even and odd numbers. Maybe all of the products end in the same digit. Take a look!

Multiplication Table

×	1	2	3	4	5	6	7	8	9	10
1	1	2	3	4	5	6	7	8	9	10
2	2	4	6	8	10	12	14	16	18	20
3	3	6	9	12	15	18	21	24	27	30
4	4	8	12	16	20	24	28	32	36	40
5	5	10	15	20	25	30	35	40	45	50
6	6	12	18	24	30	36	42	48	54	60
7	7	14	21	28	35	42	49	56	63	70
8	8	16	24	32	40	48	56	64	72	80
9	9	18	27	36	45	54	63	72	81	90
10	10	20	30	40	50	60	70	80	90	100

Select a GREEN crayon	Shade column 2. What pattern do you see? ________	Shade column 4. What pattern do you see? ________	Compare columns 2 and 4. What do you notice? ________
Select a BLUE crayon	Shade row 5. What pattern do you see? ________	Shade row 10. What pattern do you see? ________	Compare rows 5 and 10. How are they the same? How are they different? ________
Select a YELLOW crayon	Shade column 3. What pattern do you see? ________	Shade column 6. What pattern do you see? ________	Compare columns 3 and 6. How are they the same? How are they different? ________
Select a RED crayon	Shade row 7. Are the products all odd? All even? In a pattern? ________	Shade row 8. Are the products all odd? All even? In a pattern? ________	Is there a row that has only odd products? ________

PRACTICE: Now you try

Solve the problems. Is the product even or odd? Were the factors even or odd?

1. 9 × 9 = ___
   odd × odd = odd
2. 4 × 3 = ___
   ___ × ___ = ___
3. 2 × 10 = ___
   ___ × ___ = ___
4. 7 × 4 = ___
   ___ × ___ = ___

5. Think about the patterns for adding even and odd numbers. How is multiplying similar to adding? How is it different?

____

Jake was practicing writing his multiplication facts for 7. His teacher noticed two errors. What two errors did Jake make while practicing his 7 facts? Use patterns to help find the mistakes and correct them. Show your work and explain your thinking on a piece of paper.

7 14 21 27 35 42 49 56 64 70

ACE IT TIME!

	yes	no
Did you underline the question in the word problem?		
Did you circle the numbers or number words?		
Did you box the clue words that tell you what operation to use?		
Did you use a picture to show your thinking?		
Did you label your numbers and your picture?		
Did you explain your thinking and use math vocabulary words in your explanation?		

Math Vocabulary

factor
product
odd
even

Math on the Move

Have an adult or a friend lend you a random number of coins. Place the coins into groups of two. Make as many groups of two as you can. Stop when you can't make any more groups. If every group has two coins, you have an even number of coins. If there is one group with only one coin, you have an odd number. What does this mean for multiplication? When you multiply by 2, will you always have an even number?

Multiplying with 10s

FOLLOWING THE OBJECTIVE
You will use place value understanding to multiply by tens.

LEARN IT: Remember your multiples of 10: 20, 30, 40, 50, 60, 70, 80, 90, and 100. There are many strategies you can use to multiply with these numbers.

Use Patterns

When multiplying by 10, there is a pattern: $1 \times 10 = 10$, $2 \times 10 = 20$, $3 \times 10 = 30$. When you multiply by tens, the place value of the other factor increases.

This works for numbers other than 10.

$$\mathbf{40 \times 2 = 80}$$
$$\mathbf{50 \times 2 = 100}$$
$$\mathbf{60 \times 2 = 120}$$

think!
$4 \times 2 = 8$.
Multiplying by a ten (40) follows the pattern.

Use Place Value

The number 50 has 5 tens and 0 ones. Use this to multiply.

20×7	5×70	40×5
= 2 (tens) × 7	= 5 × 7 (tens)	= 4 (tens) × 5
= 14 (tens)	= 35 (tens)	= 20 (tens)
= 140	= 350	= 200

Use Properties

Remember your multiplication properties.

$$30 \times 4 =$$
$$30 \times 4 = (3 \times 10) \times 4$$
$$(3 \times 10 \times 4) = (3 \times 4) \times 10$$
$$(3 \times 4) \times 10 = 12 \times 10 = \mathbf{120}$$

think!
Which properties are we using?

PRACTICE: Now you try

Use place value to find the product.

Sample:	1. 30×6	2. 50×9	3. 60×4
40×9			
$= 4$ (tens) $\times 9$	$=$ ______ $\times$ ____	$=$ ______ $\times$ ____	$=$ ______ $\times$ ____
$= 36$ (tens)	$=$ ______	$=$ ______	$=$ ______
$= 360$	$=$ ______	$=$ ______	$=$ ______

Lori's mother found notebooks on sale at the store. Each notebook had 70 pages. Lori needed 5 notebooks to start school with. How many pages of paper did she have? Show your work and explain your thinking on a piece of paper.

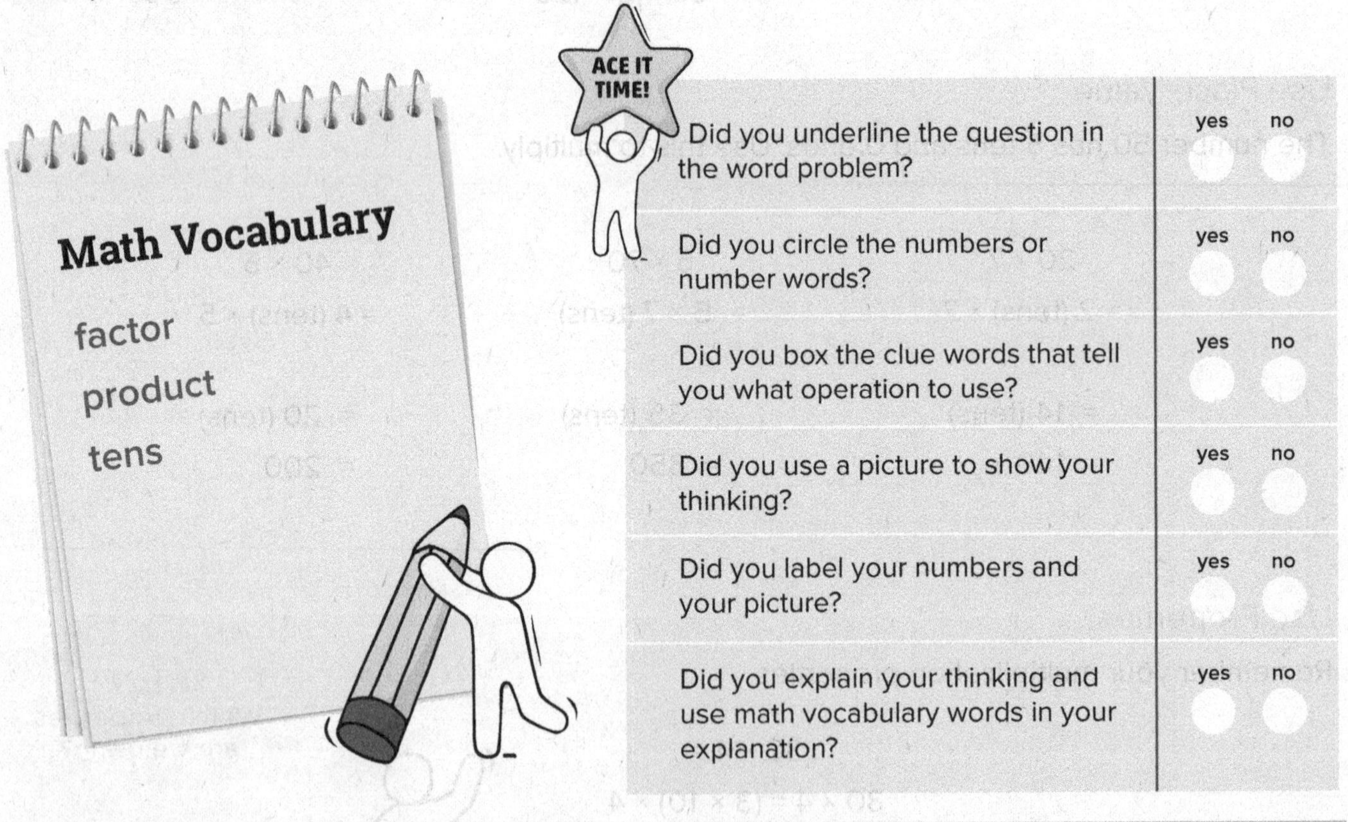

	yes	no
Did you underline the question in the word problem?		
Did you circle the numbers or number words?		
Did you box the clue words that tell you what operation to use?		
Did you use a picture to show your thinking?		
Did you label your numbers and your picture?		
Did you explain your thinking and use math vocabulary words in your explanation?		

Math on the Move

Dimes are great tools to practice this skill with. Find some dimes and group them together. Make a group of 4 dimes. Next, double the group. How many dimes do you have now? How many cents is that in total? Triple or quadruple the dimes, and so forth.

Stop and think about what you have learned.

Congratulations! You've finished the lessons for this unit. This means you know how adding and multiplying are alike. You've learned about properties of multiplication. You've practiced different ways to make multiplying easier. You can even multiply numbers up to 100.

Now it's time to prove your multiplication skills. Solve the problems below! Use all of the methods you have learned.

Activity Section

Use repeated addition and find the product for each equal group below:

1. 6 groups of 6

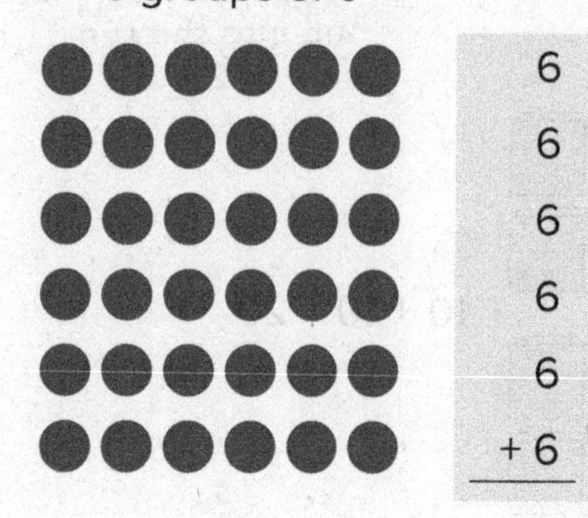

$6 \times 6 =$ ________

2. 3 groups of 2

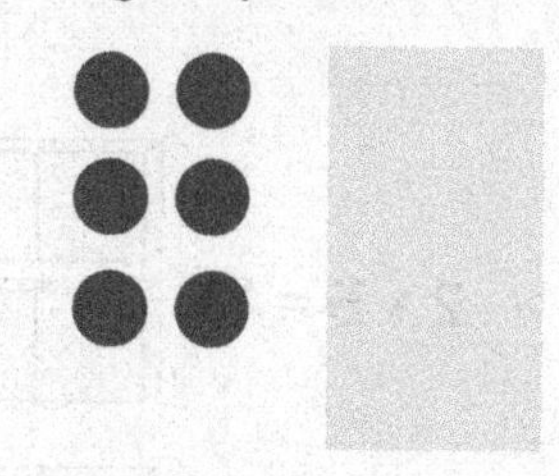

$3 \times 2 =$ ________

3. 3 groups of 4

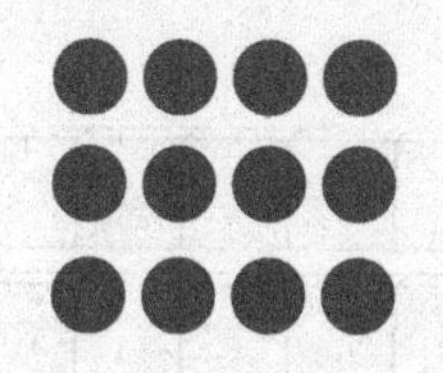

$3 \times 4 =$ ________

Complete the number sequence by skip-counting.

4. 5, 15, ______, ______, ______, ______

5. ______, 20, 30, ______, ______, ______

6. ______, ______, 300, 400, ______, ______

7. 10, 12, ______, ______, ______, ______

8. 32, 36, ______, ______, ______, ______

9. ______, 6, 9, ______, ______, ______

Use the Commutative Property of Multiplication and write the related multiplication fact.

Solve	Related Multiplication Fact
10. 4 × 7 = ______	______
11. 6 × 9 = ______	______
12. 5 × 3 = ______	______
13. 8 × 7 = ______	______

Use the Distributive Property of Multiplication to break apart the larger array into two smaller arrays. Add to find the sum of the smaller arrays:

14. 4 rows of 5 is the same as 4 × 5 = ______

think!
I could break 4 up into the sum of 2 + 2.

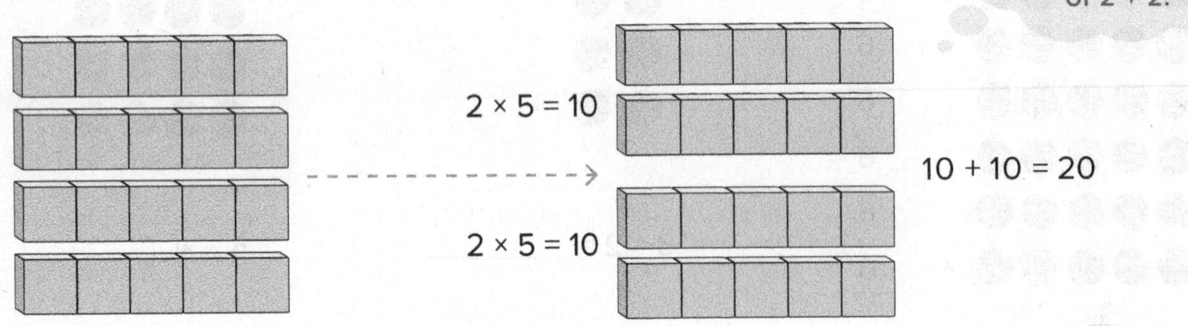

15. 6 rows of 7 is the same as 6 × 7 = ______

think!
I could break 6 up into the sum of 3 + 3.

Using the Associative Property, write another way to group factors together. Then find the product.

Sample: (5 × 3) × 2 = 5 × (3 × 2) = 5 × 6 = 30	**16.** 3 × (4 × 3) = ______ = ______ = ______	**17.** (5 × 5) × 2 = ______ = ______ = ______	**18.** (7 × 4) × 2 = ______ = ______ = ______

Choose a method of multiplying by multiples of 10 to find the product.

19. 30 × 5	**20.** 40 × 3	**21.** 80 × 2	**22.** 50 × 9

Solve these multiplication facts.

23. 5 × 4 = ______	**24.** 4 × 10 = ______	**25.** 10 × 6 = ______	**26.** 7 × 2 = ______
27. 6 × 7 = ______	**28.** 9 × 5 = ______	**29.** 2 × 5 = ______	**30.** 9 × 0 = ______
31. 6 × 8 = ______	**32.** 3 × 5 = ______	**33.** 0 × 0 = ______	**34.** 8 × 3 = ______

UNDERSTAND

Understand the meaning of what you have learned and apply your knowledge.

You can use properties of operations to multiply numbers.

Activity Section

Mr. Black wanted his students to understand the Distributive Property of Multiplication. He gave his students this array and challenged them to find at least six different ways they could break the array apart and solve it. He reminded them to:

1. Only break apart or distribute one factor at a time.
2. Make sure the other factor stays the same.
3. Double check their work.

There would be a ten-point bonus for any student who found more than six ways to meet his challenge.

think!

Eight rows of nine is the same as $8 \times 9 =$ _______

Mr. Black's Array

●●●●●●●●●
●●●●●●●●●
●●●●●●●●●
●●●●●●●●●
●●●●●●●●●
●●●●●●●●●
●●●●●●●●●
●●●●●●●●●

Six ways I could break apart this array to solve it are:

1.	**2.**
3.	**4.**
5.	**6.**

Bonus: Was there another way?

Discover how you can apply the information you have learned.

You can identify patterns in a table.

Activity Section

Describe a pattern for the table and complete the table.

Laura's teacher suggested that she could use an "organized list" or "table" to help her solve multiplication problems like this one:

"Dalton bought five boxes of granola bars. Each box had eight bars. How many granola bars did Dalton buy?"

Box	1	2	3	4	5
Granola bars	8	16	24	32	40

Laura wanted to use this strategy to solve a problem she had for homework. Look at the table Laura created for her problem.

Packages	1	2	3	4	5
Pencils	4	8	12	____	____

1. Describe a pattern for Laura's table and complete the table. ______________________

____

____

2. Write a word problem that could have been the question Laura was trying to solve for her teacher. ______________________

____

____

3. Use Laura's strategy to solve the problem below.

Michael noticed the mealworms he was observing each had six legs. There were five mealworms. How would he complete the chart?

Mealworms	1	2	3	4	5
Legs					

4. How many mealworm legs did Michael observe? ______________________

Division Concepts

Making Equal Groups

FOLLOWING THE OBJECTIVE
You will understand that a quotient represents the size of a group or the number of equal groups made by dividing.

LEARN IT: Dividing lets you separate numbers or objects into equal groups or shares.

Example: James wants to place 12 baseball cards evenly onto 3 pages in his card collection album. How could James place the 12 cards onto the 3 pages?

Now evenly share these 12 cards on the 3 pages of the album. $12 \div 3$

Page 1

Page 2

Page 3

James puts four cards on each page. $12 \div 3 = 4$. The number 4 is the answer to the division problem. It's called a **quotient**. This quotient shows the size of each group (4 cards).

think!
You divided 12 cards by a number of groups (3 pages) and got a group size (4 cards). Can you divide by size to get a number of groups? What happens when you divide 12 cards into groups of 4 cards ($12 \div 4$)?

PRACTICE: Now you try

Use drawings to divide the objects into equal groups.

Number of Items	Number of Groups	Quick Draw and Divide
10 squares	2 groups	$10 \div 2 = 5$
1. 8 triangles	4 groups	$8 \div 4 =$ _______

Michael has 28 marbles. He puts them into 4 bags. The same number of marbles is placed in each bag. How many are in each bag? Show your work and explain your thinking on a piece of paper.

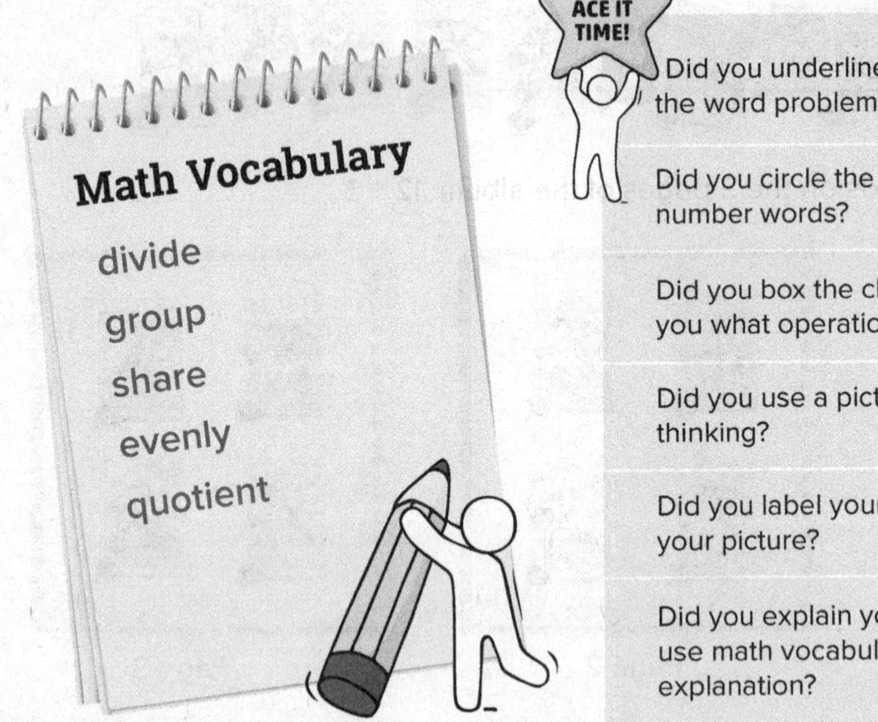

	yes	no
Did you underline the question in the word problem?	◯	◯
Did you circle the numbers or number words?	◯	◯
Did you box the clue words that tell you what operation to use?	◯	◯
Did you use a picture to show your thinking?	◯	◯
Did you label your numbers and your picture?	◯	◯
Did you explain your thinking and use math vocabulary words in your explanation?	◯	◯

Math on the Move

Sharing makes dividing real. You can share food evenly with others. Share a box of raisins, a bowl of grapes, or other shareable foods with 2 people. Make sure everyone gets the same amount! Try sharing with 3 or 4 people the next time.

Connecting Subtraction and Division

FOLLOWING THE OBJECTIVE
You will divide numbers up to 100 by using subtraction.

LEARN IT: Remember what you learned about multiplying. Multiplying is a fast way to do repeated addition. Dividing is the opposite. Dividing is a fast way to do repeated subtraction.

Example: Look at James's card collection album again. Divide 12 cards evenly on 3 pages.

How do you model this with subtraction?

Page 1

Page 2

Page 3

Start with 12. Subtract **3**.

This is the same as placing one card on each page.

12 – **3** = 9

Repeat until there are none left.

9 – **3** = 6

6 – **3** = 3

3 – **3** = 0

think!
When you multiply 2 × 5, you add 2 five times. When you divide 10 by 2, you subtract 2 five times.

PRACTICE: Now you try

1. $15 \div 5 =$ ______	**2.** $21 \div 7 =$ ______
3. $8 \div 2 =$ ______	**4.** $30 \div 10 =$ ______

Nylah has 36 books. She has a bookshelf with 4 shelves on it. If Nylah puts the same number of books on each shelf, how many books will be on each shelf? Show your work and explain your thinking on a piece of paper.

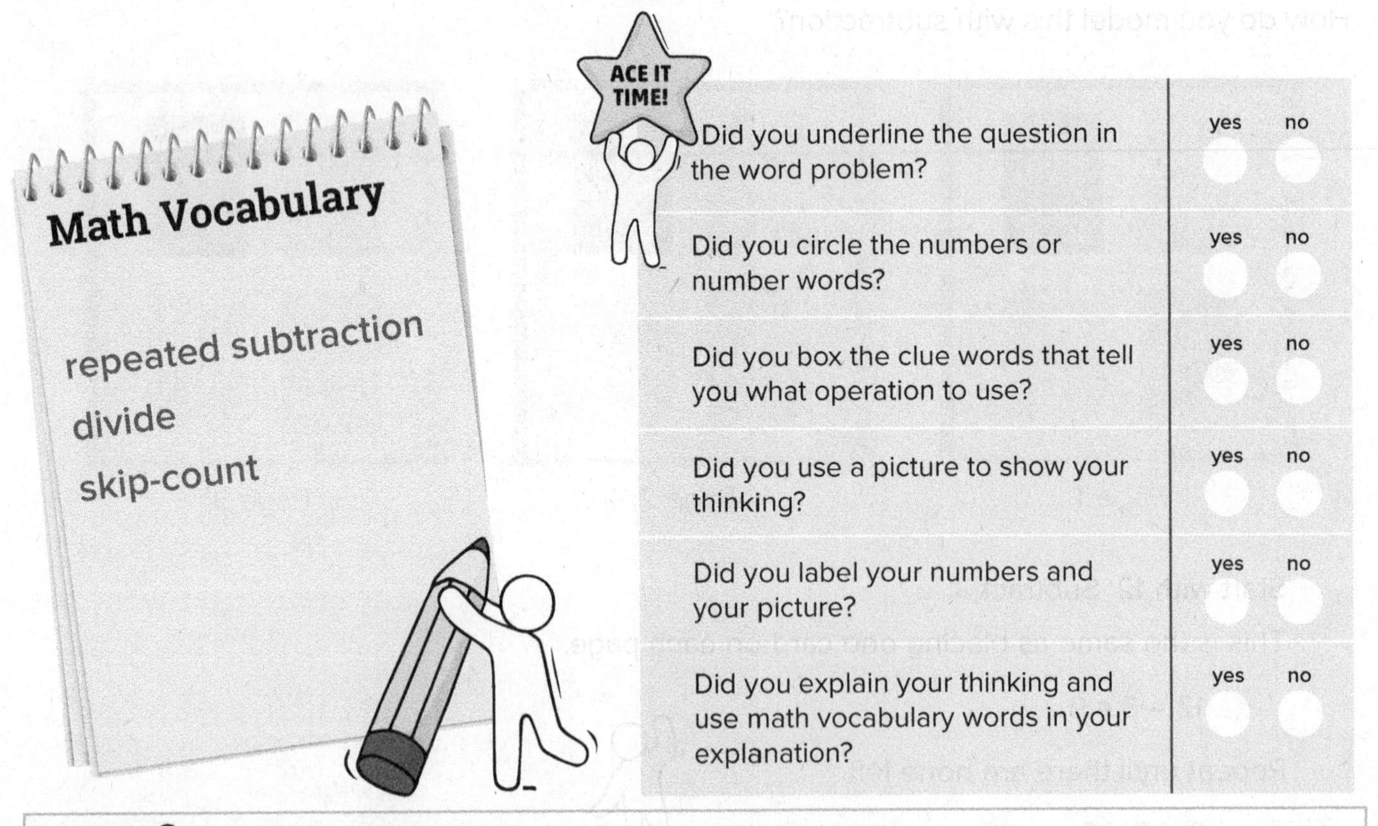

Math on the Move

The division process begins with the total or whole amount. This number is known as the *dividend*. To help you remember the name, think of this as the number we will "**divide – n – to**." The dividend is the number you begin to subtract, or skip backwards, from on a number line.

Dividing with Arrays

FOLLOWING THE OBJECTIVE
You will use arrays to divide numbers up to 100.

LEARN IT: One way of modeling division is to draw arrays. This is similar to using arrays to multiply.

Example: Divide 15 by 5.

Start with an array of 15.	Break apart into groups of 5.	How many groups of 5 are there?
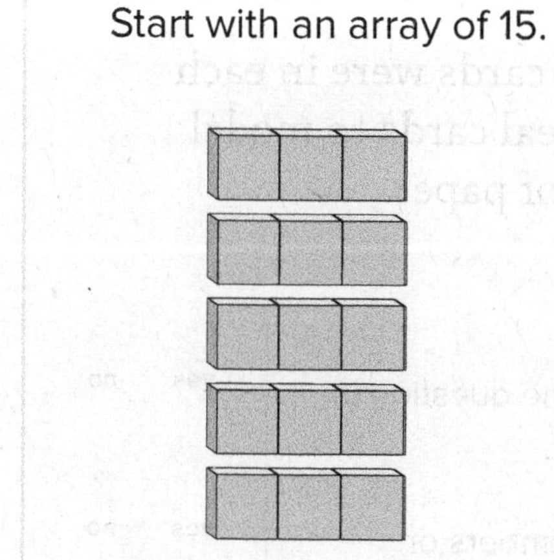		There are 3 groups of 5. **15 ÷ 5 = 3**

PRACTICE: Now you try

Solve by breaking apart each array into groups. Solve the division fact.

1. 18 ÷ 3 = _______ 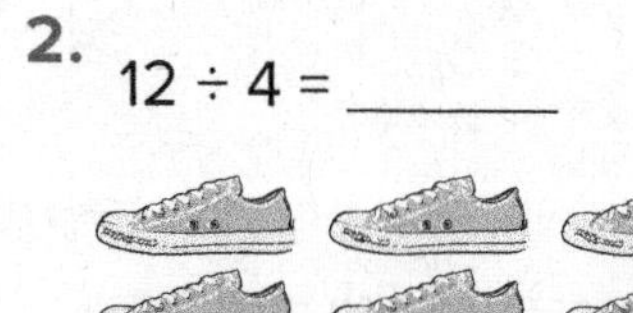	
2. 12 ÷ 4 = _______ 	

3. 25 ÷ 5 = ______

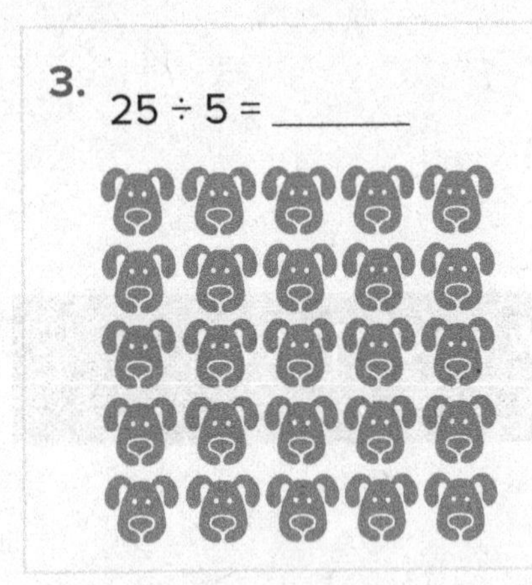

Jose made a square-shaped array with 36 cards. How many cards were in each row? *Hint:* A square has sides of equal length. You can use real cards to model this! Show your work and explain your thinking on a piece of paper.

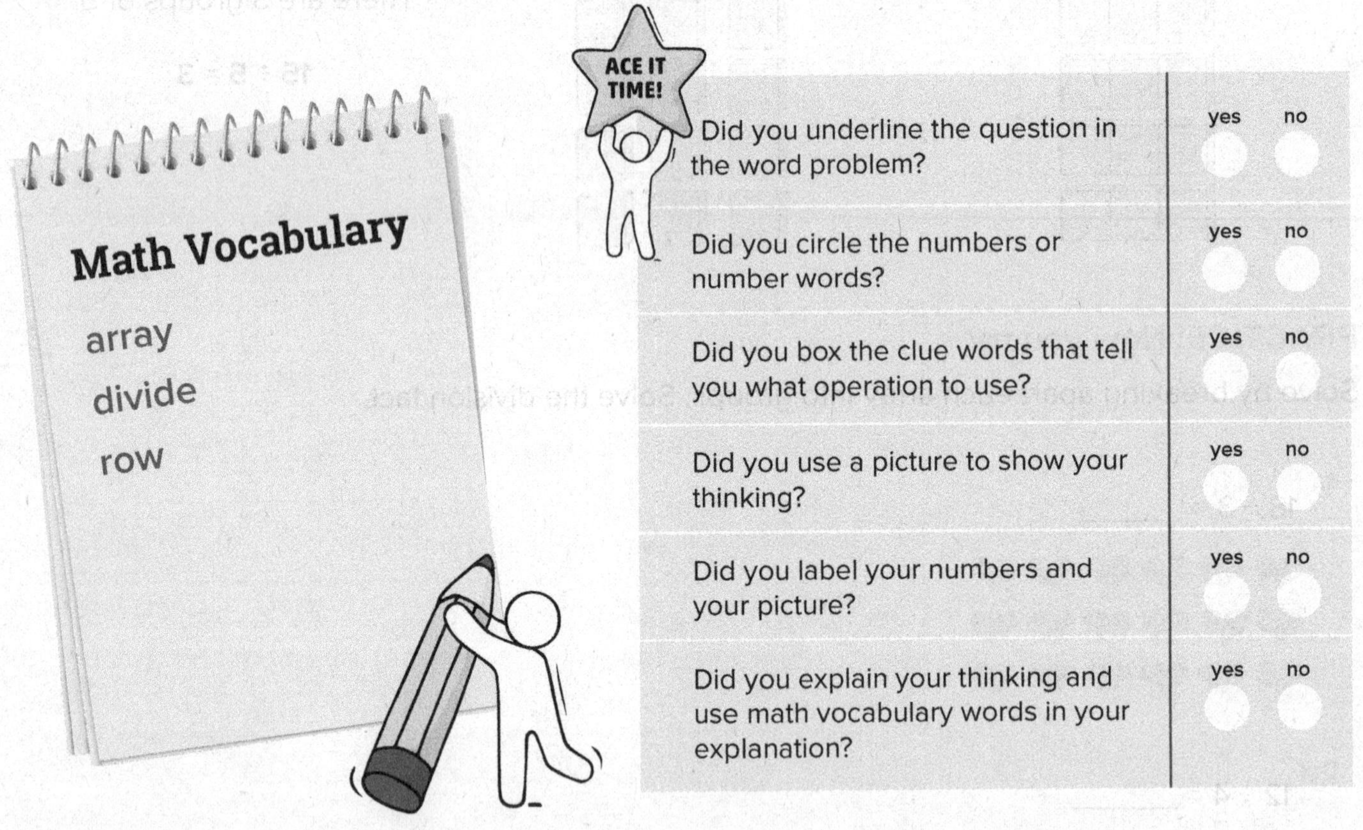

Math on the Move

Use sheets of graph/grid paper to make arrays. Cut out a square- or rectangle-shaped array from the grid paper. Count the number of blocks in your array. Next, cut it into equal groups. Write down the division fact you've just modeled. For example, make an array with 20 blocks. Separate the array into five groups. There will be four blocks in each group. This shows 20 ÷ 4 = 5.

Connecting Multiplication and Division

FOLLOWING THE OBJECTIVE
You will understand the relationship between multiplication and division.

LEARN IT: Dividing is the opposite of multiplying.

Example: Look at the fact family of 3, 4, and 12. The same array can represent a multiplication fact and its related division fact.

Multiplication	3 groups of 4	4 groups of 3
3 × 4 = 12 4 × 3 = 12		
Division	12 ÷ 4 = 3	12 ÷ 3 = 4
12 ÷ 4 = 3 12 ÷ 3 = 4		

PRACTICE: Now you try

Complete the equations.

1.

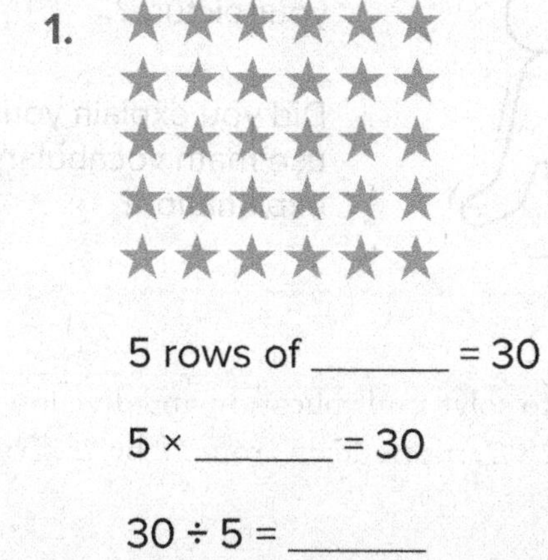

5 rows of _______ = 30

5 × _______ = 30

30 ÷ 5 = _______

2.

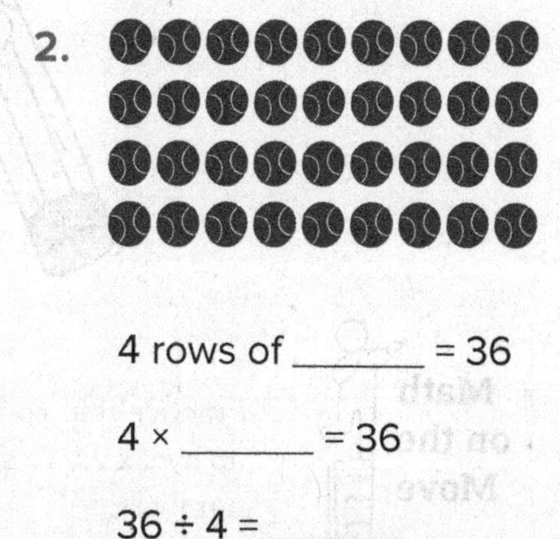

4 rows of _______ = 36

4 × _______ = 36

36 ÷ 4 = _______

Write all multiplication and division facts for the number family.

Sample:	3.	4.
3, 4, 12	5, 10, 50	3, 6, 18
3 × 4 = 12	____________	____________
4 × 3 = 12	____________	____________
12 ÷ 4 = 3	____________	____________
12 ÷ 3 = 4	____________	

Marian went to the county fair. Each ride on the Ferris wheel cost 4 tickets. She used a total of 24 tickets to ride the Ferris wheel. How many times did Marian ride the Ferris wheel? Show your work and explain your thinking on a piece of paper.

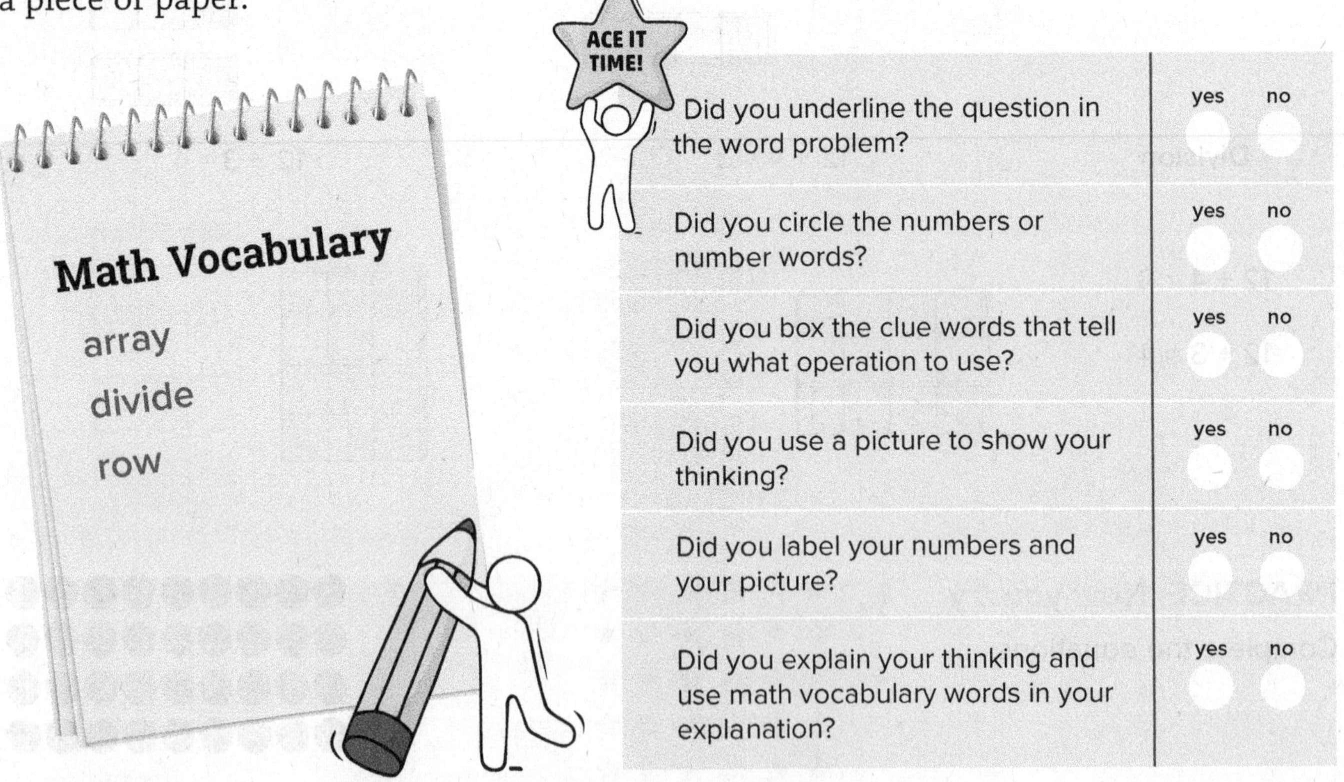

	yes	no
Did you underline the question in the word problem?	○	○
Did you circle the numbers or number words?	○	○
Did you box the clue words that tell you what operation to use?	○	○
Did you use a picture to show your thinking?	○	○
Did you label your numbers and your picture?	○	○
Did you explain your thinking and use math vocabulary words in your explanation?	○	○

Math on the Move

Practice skip-counting to solve multiplication and division problems. To solve 2 × 6, count up by six 2s. To solve 12 ÷ 2, count down from 12 by 2s. Try skip-counting for other numbers.

Multiplying and Dividing by 0 and 1

FOLLOWING THE OBJECTIVE
You will multiply and divide numbers by 0 and 1.

LEARN IT: There are special rules for multiplying and dividing by 0 and 1.

Multiplying is a fast way to add numbers.
Multiplying $3 \times 2 = 6$ is the same as adding $3 + 3 = 6$. You are adding two 3s.

Dividing is a fast way to subtract to get to zero.
Solving $6 \div 3 = 2$ is the same as subtracting $6 - 3 - 3 = 0$. You are subtracting two 3s.

What happens when you multiply or divide by 0s and 1s?

$5 \times 0 = 0$ $0 \div 5 = 0$

$5 \times 1 = 5$ $5 \div 1 = 5$

think!
Why didn't we write $5 \div 0 = 0$? Dividing is a fast way of subtracting to zero. What happens when you subtract $5 - 0 - 0 - 0 - 0 - 0 \ldots$?

5×0 means there are zero groups of five. There is nothing to draw! **5 × 0 = 0**	$0 \div 5$ means you draw an array of zero. Break apart into groups of five. **0 ÷ 5 = 0** *Wait!* If there is nothing to draw, can you break apart into groups? No.
5×1 means there is one group of five. 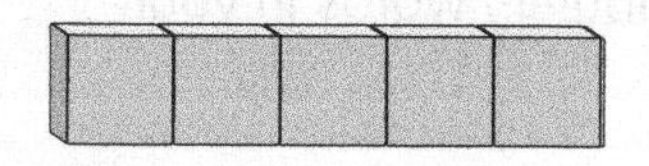 **5 × 1 = 5**	$5 \div 1$ means you draw an array of 5. Break it into equal groups of one. 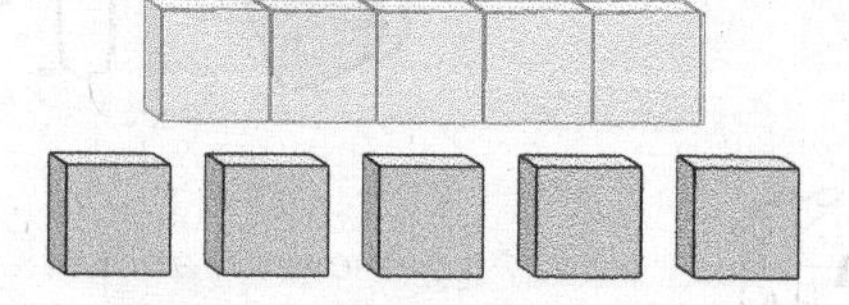 **5 ÷ 1 = 5**

These arrays show the Zero Property and the Identity Property.

Operation	Zero Property	Identity Property
Multiplication	Any number multiplied by zero equals zero. $5 \times 0 = 0$ and $0 \times 5 = 0$	A number multiplied by one keeps its identity. $5 \times 1 = 5$ and $5 \times 1 = 5$
Division	Zero divided by any number equals zero. $0 \div 5 = 0$	Any number divided by one keeps its identity. $5 \div 1 = 5$ Any number divided by itself equals 1. $5 \div 5 = 1$

think! Can you divide zero by zero?

PRACTICE: Now you try

Find the product or quotient. **1.** $9 \div 1 =$ ________ **2.** $4 \times 0 =$ ________

Complete the multiplication or division sentence. **3.** ________ $\times 19 = 0$

Mary has 6 vases. There are no flowers in any of her vases. Jane has 6 vases. There is one flower in each of her vases. Mary says they have the same number of flowers. Is she correct? Show your work and explain your thinking on a piece of paper.

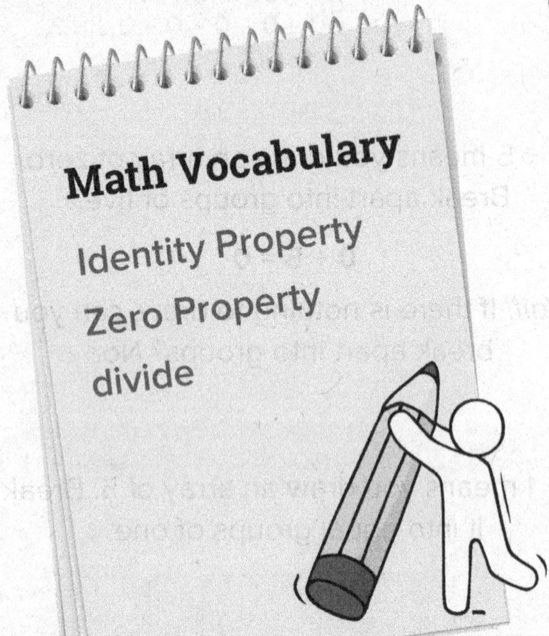

ACE IT TIME!

	yes	no
Did you underline the question in the word problem?	◯	◯
Did you circle the numbers or number words?	◯	◯
Did you box the clue words that tell you what operation to use?	◯	◯
Did you use a picture to show your thinking?	◯	◯
Did you label your numbers and your picture?	◯	◯
Did you explain your thinking and use math vocabulary words in your explanation?	◯	◯

Math on the Move

Play a game with an adult or a friend. Think of a very large number. Multiply it by 1 or 0. Ask, "Did you know . . . ?" For example, multiply infinity by zero. Say, "Did you know infinity times zero equals zero?" See how many large numbers you can multiply or divide correctly!

Division Facts

FOLLOWING THE OBJECTIVE
You will divide numbers under 100.

LEARN IT: In higher grades, you will divide large numbers. It will be easier to do this if you memorize division facts for smaller numbers. Make flashcards for the division facts listed below. Practice these division facts every day until you remember them all.

zero	one	two	three	four	five
0 ÷ 1 = 0	1 ÷ 1 = 1	2 ÷ 2 = 1	3 ÷ 3 = 1	4 ÷ 4 = 1	5 ÷ 5 = 1
0 ÷ 2 = 0	2 ÷ 1 = 2	4 ÷ 2 = 2	6 ÷ 3 = 2	8 ÷ 4 = 2	10 ÷ 5 = 2
0 ÷ 3 = 0	3 ÷ 1 = 3	6 ÷ 2 = 3	9 ÷ 3 = 3	12 ÷ 4 = 3	15 ÷ 5 = 3
0 ÷ 4 = 0	4 ÷ 1 = 4	8 ÷ 2 = 4	12 ÷ 3 = 4	16 ÷ 4 = 4	20 ÷ 5 = 4
0 ÷ 5 = 0	5 ÷ 1 = 5	10 ÷ 2 = 5	15 ÷ 3 = 5	20 ÷ 4 = 5	25 ÷ 5 = 5
0 ÷ 6 = 0	6 ÷ 1 = 6	12 ÷ 2 = 6	18 ÷ 3 = 6	24 ÷ 4 = 6	30 ÷ 5 = 6
0 ÷ 7 = 0	7 ÷ 1 = 7	14 ÷ 2 = 7	21 ÷ 3 = 7	28 ÷ 4 = 7	35 ÷ 5 = 7
0 ÷ 8 = 0	8 ÷ 1 = 8	16 ÷ 2 = 8	24 ÷ 3 = 8	32 ÷ 4 = 8	40 ÷ 5 = 8
0 ÷ 9 = 0	9 ÷ 1 = 9	18 ÷ 2 = 9	27 ÷ 3 = 9	36 ÷ 4 = 9	45 ÷ 5 = 9
0 ÷ 10 = 0	10 ÷ 1 = 10	20 ÷ 2 = 10	30 ÷ 3 = 10	40 ÷ 4 = 10	50 ÷ 5 = 10

six	seven	eight	nine	ten
6 ÷ 6 = 1	7 ÷ 7 = 1	8 ÷ 8 = 1	9 ÷ 9 = 1	10 ÷ 10 = 1
12 ÷ 6 = 2	14 ÷ 7 = 2	16 ÷ 8 = 2	18 ÷ 9 = 2	20 ÷ 10 = 2
18 ÷ 6 = 3	21 ÷ 7 = 3	24 ÷ 8 = 3	27 ÷ 9 = 3	30 ÷ 10 = 3
24 ÷ 6 = 4	28 ÷ 7 = 4	32 ÷ 8 = 4	36 ÷ 9 = 4	40 ÷ 10 = 4
30 ÷ 6 = 5	35 ÷ 7 = 5	40 ÷ 8 = 5	45 ÷ 9 = 5	50 ÷ 10 = 5
36 ÷ 6 = 6	42 ÷ 7 = 6	48 ÷ 8 = 6	54 ÷ 9 = 6	60 ÷ 10 = 6
42 ÷ 6 = 7	49 ÷ 7 = 7	56 ÷ 8 = 7	63 ÷ 9 = 7	70 ÷ 10 = 7
48 ÷ 6 = 8	56 ÷ 7 = 8	64 ÷ 8 = 8	72 ÷ 9 = 8	80 ÷ 10 = 8
54 ÷ 6 = 9	63 ÷ 7 = 9	72 ÷ 8 = 9	81 ÷ 9 = 9	90 ÷ 10 = 9
60 ÷ 6 = 10	70 ÷ 7 = 10	80 ÷ 8 = 10	90 ÷ 9 = 10	100 ÷ 10 = 10

PRACTICE: Now you try

Look at each picture. Count the number of groups and the number of objects in each group. Write the division fact.

Answer each division problem. Try to do this by memory!

Sample:

10 ÷ 2 = 5 and 10 ÷ 5 = 2

1.  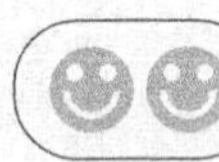 

____

The book Caylie is reading is 72 pages total. Each chapter has 8 pages. How many chapters are in Caylie's book? Show your work and explain your thinking on a piece of paper.

ACE IT TIME!

	yes	no
Did you underline the question in the word problem?	○	○
Did you circle the numbers or number words?	○	○
Did you box the clue words that tell you what operation to use?	○	○
Did you use a picture to show your thinking?	○	○
Did you label your numbers and your picture?	○	○
Did you explain your thinking and use math vocabulary words in your explanation?	○	○

Math Vocabulary

divide

quotient

Math on the Move

Rhymes can help you remember division facts. For example, "Seven and seven are doing fine, 7 × 7 is 49," or "Six and four are on the floor, 6 × 4 is 24." Make up rhymes for the facts you find difficult.

Find the Unknown Number

FOLLOWING THE OBJECTIVE
You will find the value of unknown numbers in multiplication and division equations.

LEARN IT: An equation is a number sentence that uses an equal sign to show that two amounts are equal. A symbol, a letter, or ■ can be used to represent an unknown number. Remember what you've learned about multiplying and dividing. You can rewrite multiplication equations as division to solve.

$4 \times 5 = 20$

Find Unknown Numbers in Multiplication Equations

When the First Factor Is Unknown ■ × 5 = 20

Use multiplication you know. What number times 5 is equal to 20? **4 × 5 = 20**	Divide the product by the factor you know (5). **20 ÷ 5 = ■** **20 ÷ 5 = 4**

When the Second Factor Is Unknown 4 × ■ = 20

Use multiplication you know. 4 times what number is equal to 20? **4 × 5 = 20**	Divide the product by the factor you know (4). **20 ÷ 4 = ■** **20 ÷ 4 = 5**

$20 \div 4 = 5$

Find Unknown Numbers in Division Equations

When the Dividend Is Unknown ■ ÷ 4 = 5

Use division you know. What number divided by 4 is equal to 5? **20 ÷ 4 = 5**	Multiply the quotient (5) by the divisor (4). **5 × 4 = ■** **5 × 4 = 20**

When the Divisor Is Unknown 20 ÷ ■ = 5

Use division you know. 20 divided by what number is equal to 5? **20 ÷ 4 = 5** You can also switch the quotient and divisor. **20 ÷ 5 = 4**	Use multiplication you know. 5 times what number is equal to 20? **5 × 4 = 20**

PRACTICE: Now you try

Find the unknown number. Use one of the methods shown on page 63.

1. $4 \times 7 =$ ______ **2.** $3 \times$ ______ $= 12$ **3.** ______ $\times 9 = 36$ **4.** $2 \times 6 =$ ______

5. $5 \times$ ______ $= 25$ **6.** ______ $\times 8 = 48$ **7.** $6 \times 9 =$ ______ **8.** $7 \times$ ______ $= 42$

Write a related division equation to find the unknown number.

9. $3 \times$ ▢ $= 12$ → $12 \div$ ______ $=$ ▢
▢ $=$ ______

10. $7 \times$ ▢ $= 49$ → $49 \div$ ______ $=$ ▢
▢ $=$ ______

Samantha had some cookies in a bag. She wanted to share them equally among herself and her 4 friends. Each friend got 7 cookies. How many cookies were in the bag? Show your work and explain your thinking on a piece of paper.

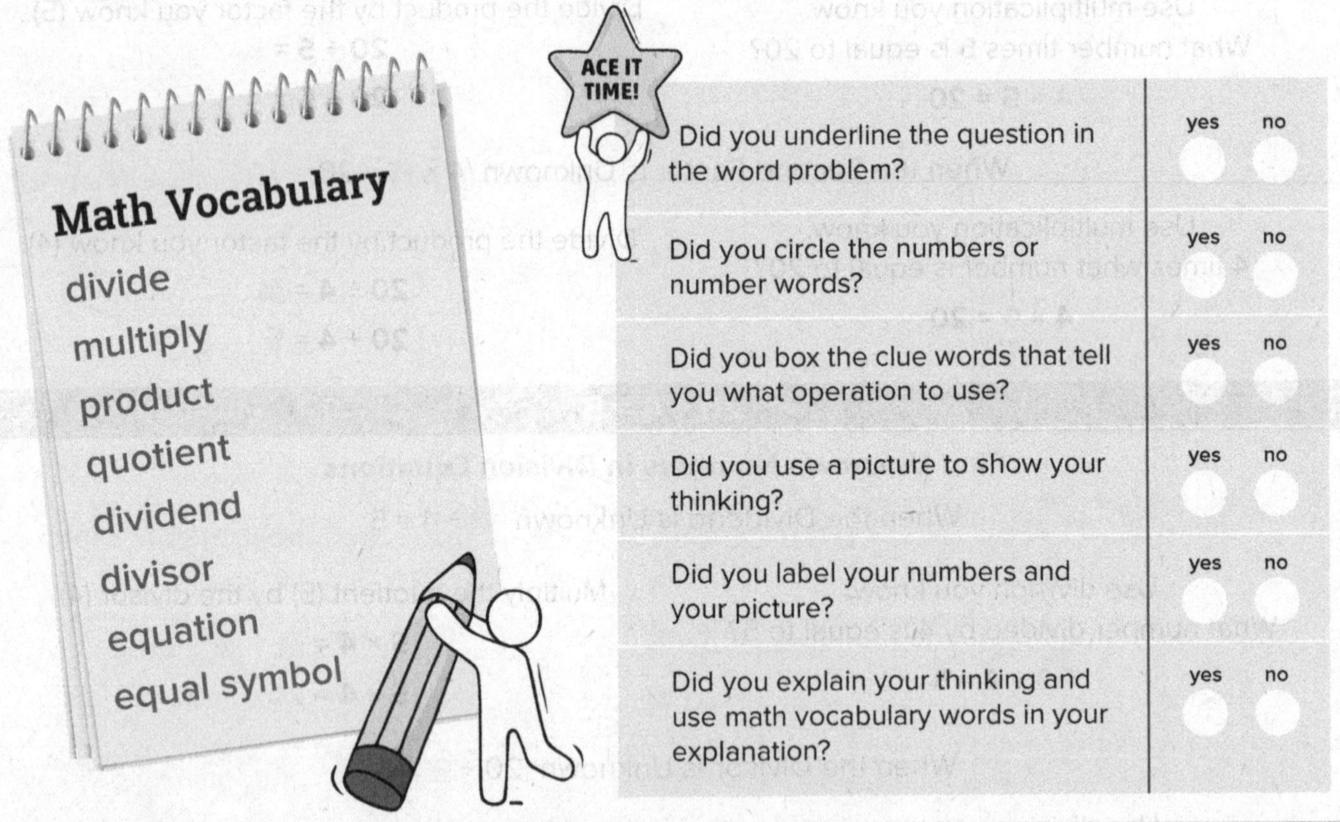

Use what you know about multiplication and division fact families to help you practice your facts. Try using opposite operations to help improve your multiplication and division skills. For example, if you want to solve $6 \div 2$, ask yourself, "What times 2 equals 6?"

Two-Step Problems and Equations

FOLLOWING THE OBJECTIVE
You will write equations to represent word problems and solve two-step problems.

LEARN IT: Story problems sometimes take two steps to solve. Two-step problems usually have an inferred question. An inferred question is one that is not written. It is drawn using evidence and reasoning. This must be solved before you can answer the written question.

Example: Byron runs 2 miles a day. His goal is to run 26 miles total. He has run for 5 days. How many more miles must Byron run before he reaches his goal?

What is the question you need to answer? **How many more miles must Byron run before he reaches his goal?**

This is a subtraction problem. To solve, subtract the number of miles Byron has run from his goal.

26 – miles Byron has already run = miles left

Does the story problem say how many miles Byron has already run? No. This is the inferred question. Step 1 is to solve this question:

Byron runs 2 miles each day. He has run for 5 days. How many miles has he run so far?

5 days × 2 miles per day = 10 miles

Step 2 is to subtract. **26 – 10 = 16**.

Byron has to run 16 more miles before he reaches his goal.

PRACTICE: Now you try

Solve the problem using two steps.

1. Roberto had 48 baseball cards. His best friend, Eddie, gave him 12 more. He wants to put all the cards in his new baseball card album. Each page in his album has space for 12 baseball cards. How many pages can he fill up with cards?

Step 1. Answer the inferred question.

Equation ______________________________

Step 2. Answer the question stated in the story problem.

Answer ______________________________

____

At Field Day, each of the 6 grades at Campbell Park Elementary entered 7 players into the games. The principal bought a package of 100 ice pops. How many ice pops are left over after each player eats one? Show your work and explain your thinking on a piece of paper.

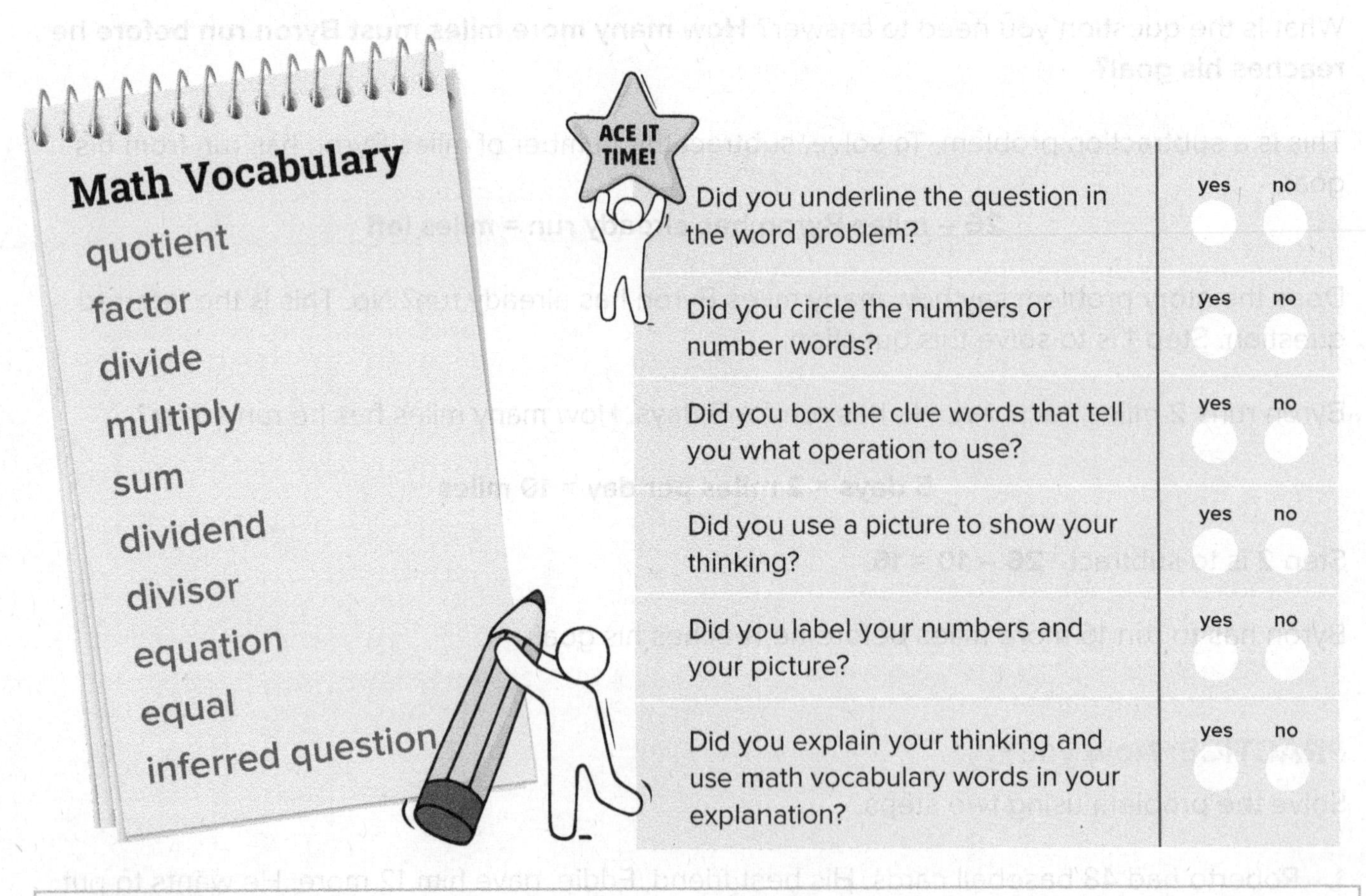

Math on the Move

Help with shopping. When you are buying groceries with your family or friends, add or multiply to determine the cost of your items. Challenge yourself to figure out the change before the cashier can.

Stop and think about what you have learned.

Congratulations! You've finished the lessons for this unit. This means you know how multiplying and dividing are related. You've learned how equal groups and arrays help you solve division problems. You've practiced different ways to make division easier. You can even divide numbers to solve two-step problems.

Now it's time to prove your division skills. Solve the problems below! Use all of the methods you have learned.

Activity Section

Create equal groups and solve.

Baseball Cards	Album Pages	Cards on Each Page
1. 15 cards	3 pages	15 ÷ 3 = ________
2. 12 cards	4 pages	12 ÷ 4 = ________
3. 16 cards	8 pages	16 ÷ 8 = ________

Complete the number sequence. Use repeated subtraction or skip-counting.

4. 80, 70, 60, _____, _____, _____, _____

5. 18, 16, 14, _____, _____, _____, _____

6. 54, _____, 36, _____, _____, _____

7. 56, 49, 42, _____, _____, _____, _____

8. _____, 90, _____, 80, 75, _____, _____

9. 28, 24, 20, _____, _____, _____, _____

Write the multiplication and division facts for each set of numbers.

Sample:

2, 4, 8

$2 \times 4 = 8$

$4 \times 2 = 8$

$8 \div 4 = 2$

$8 \div 2 = 4$

10. **3, 10, 30**

____

____

____

____

11. **6, 8, 48**

____

____

____

____

12. **6, 7, 42**

____

____

____

____

13. **18, 3, 6**

____

____

____

____

14. **5, 8, 40**

____

____

____

____

15. **54, 6, 9**

____

____

____

____

16. **8, 8, 64**

____

____

Solve these division facts.

17. 32 ÷ 4 = ______ **18.** 60 ÷ 10 = ______ **19.** 45 ÷ 5 = ______

20. 49 ÷ 7 = ______ **21.** 32 ÷ 8 = ______ **22.** 16 ÷ 4 = ______

23. 15 ÷ 3 = ______ **24.** 54 ÷ 9 = ______ **25.** 15 ÷ ______ = 3

26. ______ ÷ 4 = 6 **27.** ______ ÷ 2 = 9 **28.** 63 ÷ ______ = 9

29. 7 ÷ ______ = 1 **30.** 21 ÷ ______ = 7 **31.** 53 ÷ 1 = ______

32. 12 ÷ 6 = ______ **33.** 32 ÷ ______ = 4 **34.** 23 ÷ 1 = ______

35. 42 ÷ ______ = 7 **36.** ______ ÷ 8 = 7 **37.** 36 ÷ 6 = ______

38. 70 ÷ 10 = ______ **39.** 9 ÷ 3 = ______ **40.** 35 ÷ 5 = ______

Understand the meaning of what you have learned and apply your knowledge.

You can use multiplication and division within 100 to solve word problems.

Activity Section

Mr. Mitton gave his students 24 tiles and showed them how they could make an array with 4 rows of 6 tiles, like the one below. He wanted them to break apart the array to show different multiplication and division facts. His directions were:

1. Divide the array four different ways.
2. Each array MUST equal a total of 24 tiles.
3. The original array (4 rows of 6) may not count as one of their ways.
4. They must include both the multiplication and division facts for each array.

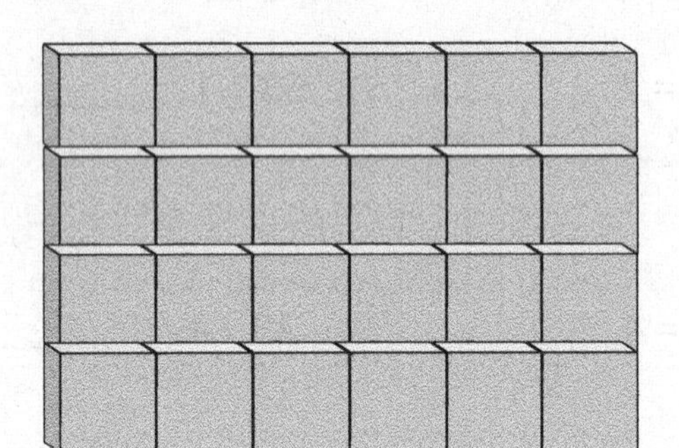

$4 \times 6 = 24$

$8 \times 3 = 24$

$12 \times 2 = 24$

$24 \times 1 = 24$

The three ways I can break up the array are:

1.

2.

3.

DISCOVER

Discover how you can apply the information you have learned.

You will use multiplication and division to determine unknown numbers. You can also solve two-step word problems.

Activity Section

Middlebrook Elementary is selling cookie dough in tubs for its fall fundraiser. Use the information in the pictures below to answer the questions.

1. Maria sold only Oatmeal Raisin Cookie Dough. She raised $60. How many tubs did she sell? Show your work.

2. Samantha's mother loves snickerdoodle cookies. She gave Samantha $20.00. What is the greatest number of Snickerdoodle Cookie Dough tubs Samantha can buy? Show your work.

3. Justin sold only Sugar Cookie Cookie Dough. He raised $24.00. Bryce sold only Snickerdoodle Cookie Dough. He also raised $24.00. Who sold more cookie dough? Show your work.

Fraction Concepts

Understanding Fractions

FOLLOWING THE OBJECTIVE:
You will practice working with equal parts of a whole.

LEARN IT: Fractional parts must be equal. A ***whole*** or group must be divided into parts of exactly the same size.

Is the shape divided into equal parts?

This shape is divided into three **equal** parts.

This shape is divided into three **unequal** parts.

The size of a fractional part is relative to the whole. If the size of the whole is different, then the size of half will be different.

Half of a king-sized candy bar is different than half of a regular-sized candy bar.

Half of this bar is $\frac{10}{20}$.

Half of this bar is $\frac{6}{12}$.

When a shape is divided into:

two parts, they are named **halves**
four parts, they are named **fourths**
six parts, they are named **sixths**
eight parts, they are named **eighths**
three parts, they are named **thirds**
five parts, they are named **fifths**
seven parts, they are named **sevenths**

What are **equal** shares?

Anthony's mom made two huge hoagies for him to share with his three brothers. What are two ways the hoagies could be divided to make equal shares?

Because there are 4 brothers total, Anthony can divide each hoagie into 4 equal parts. Each brother will get 2 equal parts out of 8. They each get $\frac{2}{8}$ of a total of two hoagies.

Because there are 4 brothers total, Anthony can divide the group of hoagies into 4 equal parts. To do this, each hoagie is cut in half. Each brother will get 1 equal part of 4. They each get $\frac{1}{4}$ of the total of two hoagies. This is the same as $\frac{1}{2}$ of one hoagie.

PRACTICE: Now you try

Write **equal** or **not equal**.

1.

2.

3.

Fill in the blank to make each statement correct.

4.

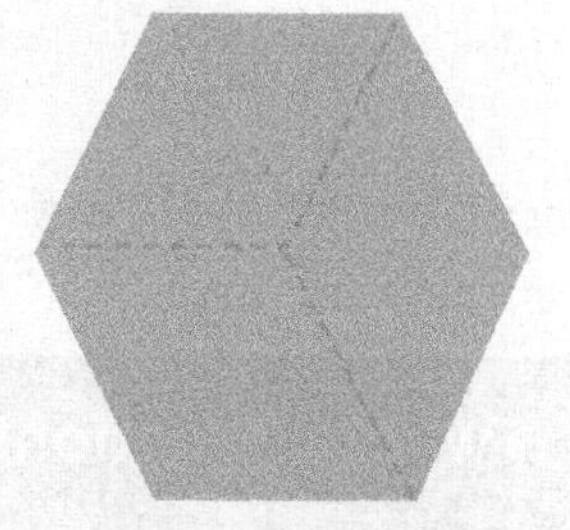

This shape is divided into ____________________

5.

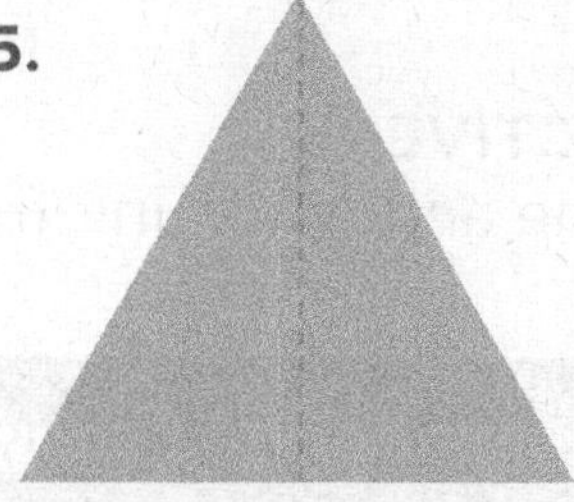

This shape is divided into ____________________

6.

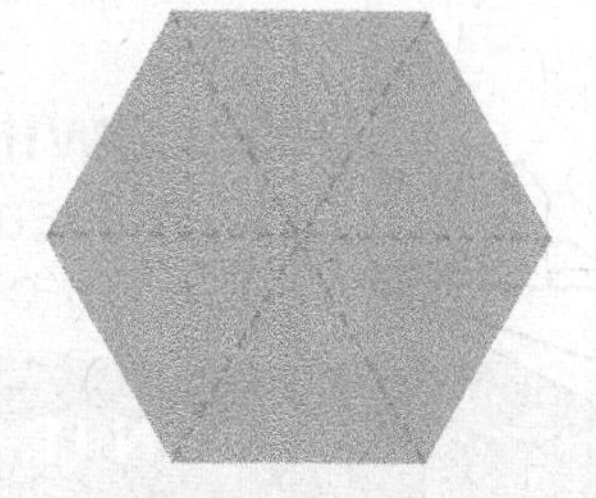

This shape is divided into ____________________

The Johnson family bought a large pizza to share with their 4 family members. The Miller family bought a medium pizza to share with their 4 family members. If each member of the family got $\frac{1}{4}$ of the whole pizza, did the members from the Johnson family and Miller family get the same amount? Show your work and explain your thinking on a piece of paper.

ACE IT TIME!

	yes	no
Did you underline the question in the word problem?		
Did you circle the numbers or number words?		
Did you box the clue words that tell you what operation to use?		
Did you use a picture to show your thinking?		
Did you label your numbers and your picture?		
Did you explain your thinking and use math vocabulary words in your explanation?		

Math Vocabulary

equal parts
fractions
whole
equal shares
fourths

Math on the Move

Practice making equal parts. You can divide food and other items around the house into equal shares. You can also locate fractions in the everyday world. Try to locate fractions in advertisements or food recipes.

Unit Fractions and Other Fractions

FOLLOWING THE OBJECTIVE
You will learn how to name and write unit fractions and other fractions made up of unit fractions.

LEARN IT: A ***fraction*** is a number that shows part of a whole. Fractions have numerators and denominators.

3 ← numerator .... number of equal parts counted
4 ← denominator ..... number of equal parts in the whole

Unit Fractions	Other Fractions
A unit fraction names 1 equal part of a whole. In a unit fraction, the numerator is always 1. For example, $\frac{1}{4}$ is a unit fraction.	Other fractions have numerators other than 1. For example, $\frac{2}{4}$ is a fraction. This is two-fourths, or two $\frac{1}{4}$ pieces. Other fractions are made of multiple unit fractions: $\frac{1}{4} + \frac{1}{4} = \frac{2}{4}$
The **numerator** means one of four equal parts. → 1 The **denominator** means that the whole is divided into four equal parts. → 4	The **numerator** means two of four equal parts. → 2 The **denominator** means that the whole is divided into four equal parts. → 4

How do you name and write fractions? Use the picture of brownies to name and write a fraction. How much of the pan of brownies has nuts?

Step 1: Identify the denominator. Count the number of equal parts in the whole pan of brownies. There are eight equal parts in the pan of brownies. So the denominator is 8.

Step 2: Identify the numerator. Count the number of equal parts with nuts. There are five equal parts with nuts. So the numerator is 5.

Numerator
Denominator $\frac{5}{8}$ Five out of eight equal brownie pieces have nuts.

PRACTICE: Now you try

Name and write the fraction. Each model equals a whole. Write the fraction for the dark gray shaded parts.

1.

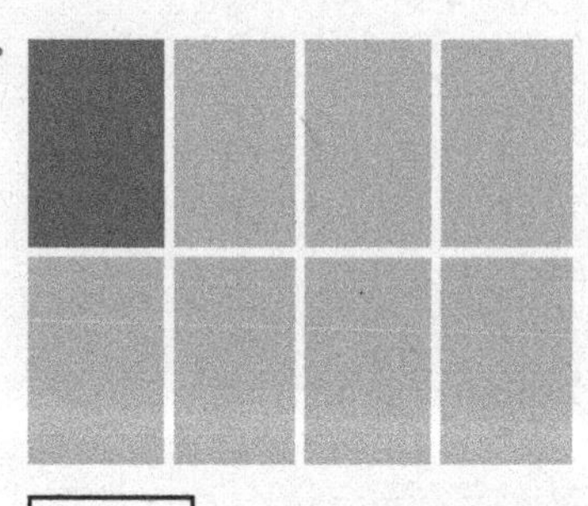

← Numerator

← Denominator

2.

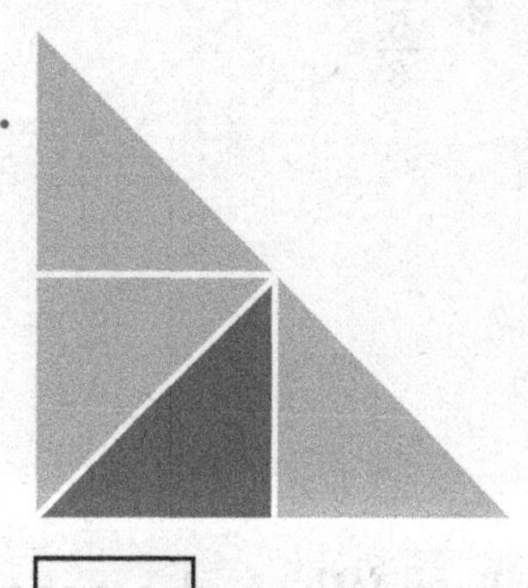

← Numerator

← Denominator

3.

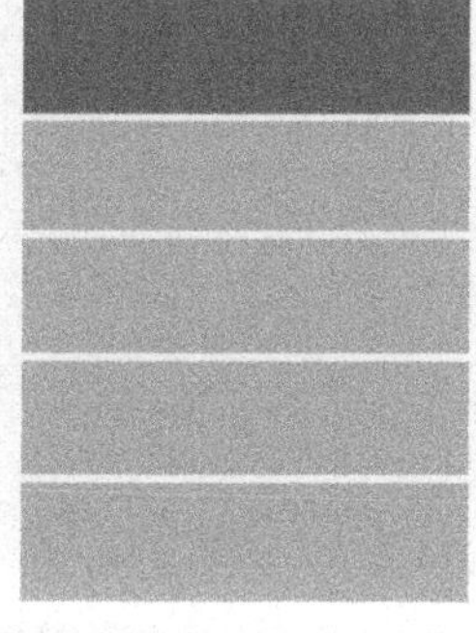

← Numerator

← Denominator

Show two different ways to divide this rectangle into fourths.

4.

Each model equals a whole. Write the fraction for the dark gray shaded parts.

5.

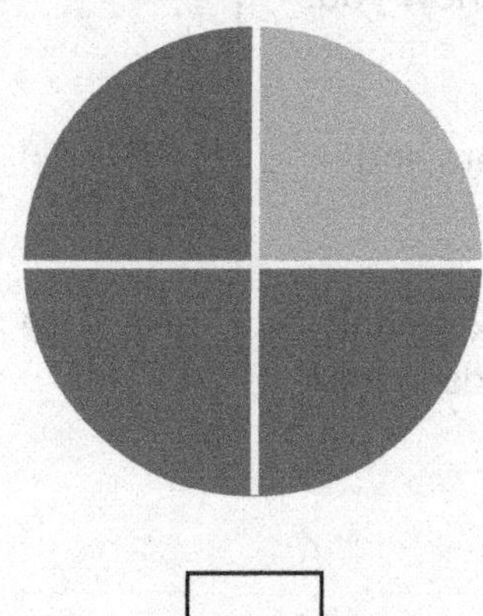

6.

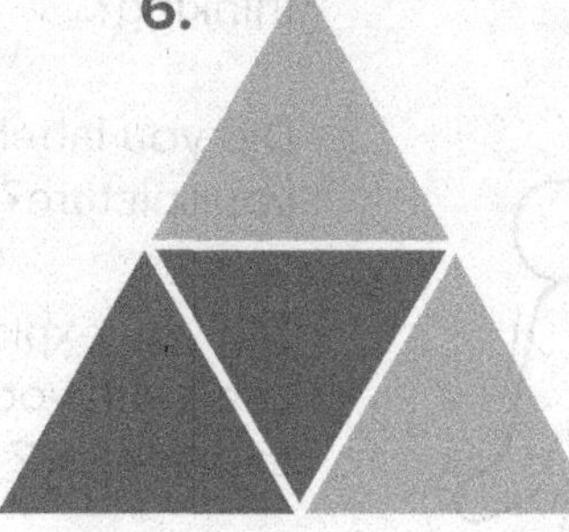

7.

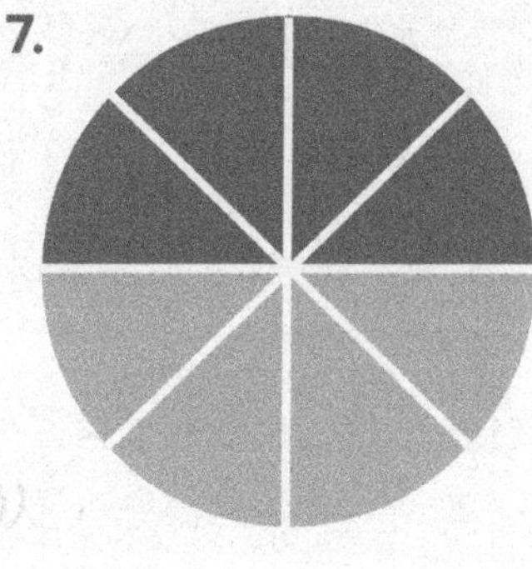

Each model represents one whole. Shade the model to represent the fraction.

8. $\frac{4}{6}$

9. $\frac{5}{8}$

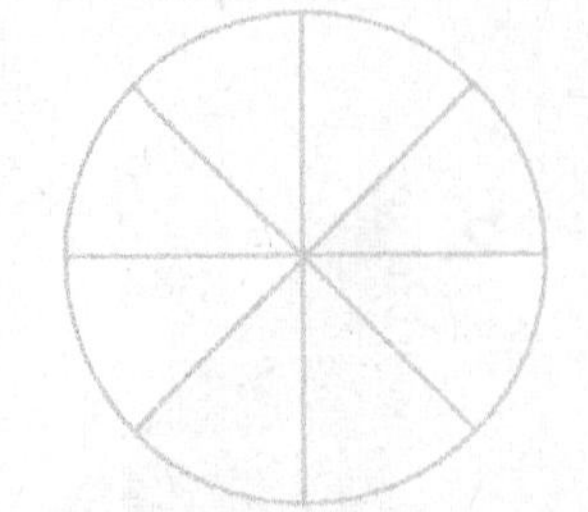

10. $\frac{3}{4}$

Lela had a candy bar in her lunch bag. When she got to lunch and pulled it out, she had already eaten $\frac{3}{5}$ of the bar. How many more pieces does Lela need to eat to finish up her candy bar? Show your work and explain your thinking on a piece of paper.

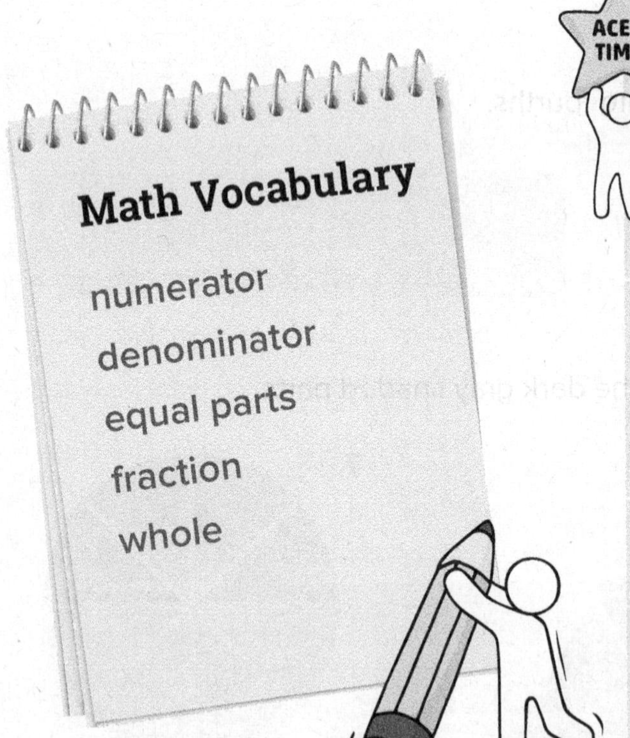

ACE IT TIME!

	yes	no
Did you underline the question in the word problem?	○	○
Did you circle the numbers or number words?	○	○
Did you box the clue words that tell you what operation to use?	○	○
Did you use a picture to show your thinking?	○	○
Did you label your numbers and your picture?	○	○
Did you explain your thinking and use math vocabulary words in your explanation?	○	○

Practice your fractions when eating and sharing food (e.g., pizza, pie, cake, brownies). Name and write fractions that match parts of the whole—parts eaten and parts left over. Make sure to focus on dividing the whole into equal parts.

Fractions on a Number Line

FOLLOWING THE OBJECTIVE
You will draw fractions on a number line.

LEARN IT: Remember that a fraction is part of a whole. To draw this on a number line, start with the whole. Then divide the whole into equal parts.

Unit Fractions on a Number Line: Locate $\frac{1}{3}$ on the number line.

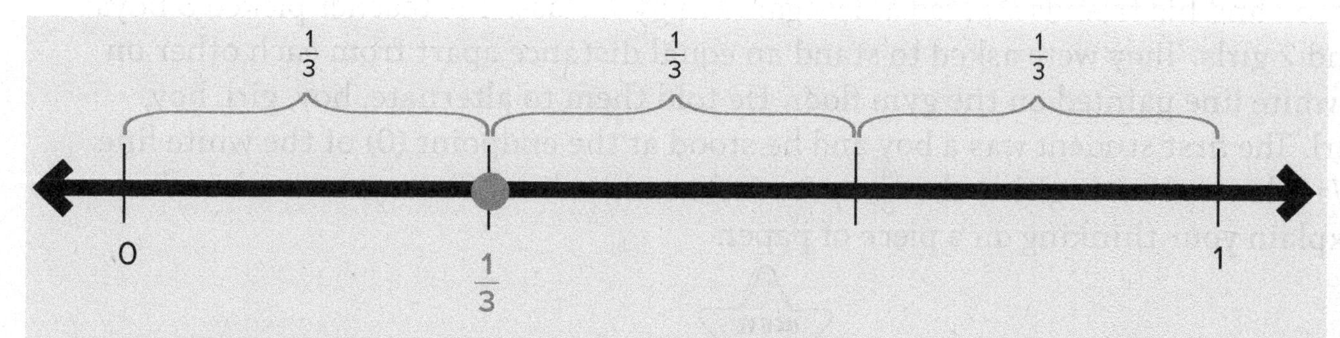

See how the distance between 0 and 1 is broken into three equal parts. Each part is equal to $\frac{1}{3}$.

Other Fractions on a Number Line: Locate $\frac{5}{4}$ on a number line.

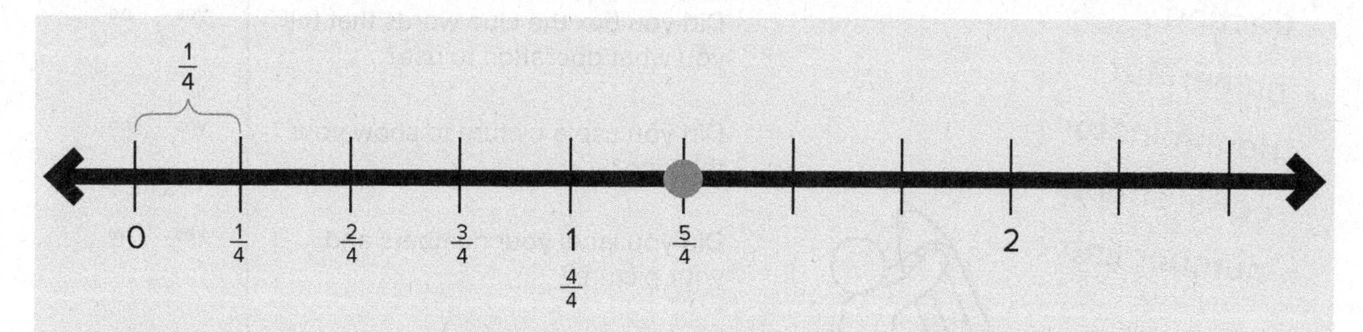

Notice how the top number (5) is greater than the bottom number (4) in $\frac{5}{4}$. This is how you know the fraction is worth more than 1. The number line will go beyond 1.

The distance between 0 and 1 is broken into four equal parts. See how the distance between 1 and 2 is also broken into four equal parts. Count five of these parts to get to $\frac{5}{4}$. This is the same as 1 whole plus $\frac{1}{4}$.

think! How many jumps do you make to get from 0 to 1? That is your denominator. Start at 1 for your numerator and label each tick on the line.

PRACTICE: Now you try

1. Fill in the missing fractions on the number line.

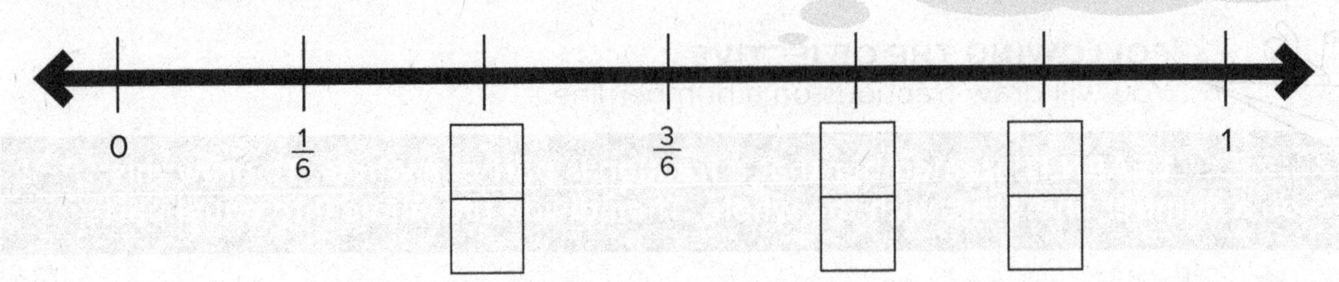

Jason and his friends started a new game in gym class. The teacher picked 3 boys and 2 girls. They were asked to stand an equal distance apart from each other on a white line painted on the gym floor. He told them to alternate, boy, girl, boy, girl. The first student was a boy and he stood at the endpoint (0) of the white line. Was there a boy or girl at the $\frac{4}{5}$ spot on that white line? Show your work and explain your thinking on a piece of paper.

ACE IT TIME!

	yes	no
Did you underline the question in the word problem?		
Did you circle the numbers or number words?		
Did you box the clue words that tell you what operation to use?		
Did you use a picture to show your thinking?		
Did you label your numbers and your picture?		
Did you explain your thinking and use math vocabulary words in your explanation?		

Math Vocabulary

- fraction
- numerator
- denominator
- equal shares
- number line

Math on the Move

Draw number lines and practice labeling them with fractions. For example, you can make a number line and divide it into thirds. Then you can label it. You can even draw a number line on the sidewalk with sidewalk chalk and label it.

Relate Fractions and Whole Numbers

FOLLOWING THE OBJECTIVE
You will understand which fractions equal one and which fractions are greater than one.

LEARN IT: You can express whole numbers as fractions. A whole number represents all the parts. This means there are 3 out of 3 thirds ($\frac{3}{3}$) or 4 out of 4 fourths ($\frac{4}{4}$). You can show this on a number line.

A number line shows the whole number 1 as a fraction.

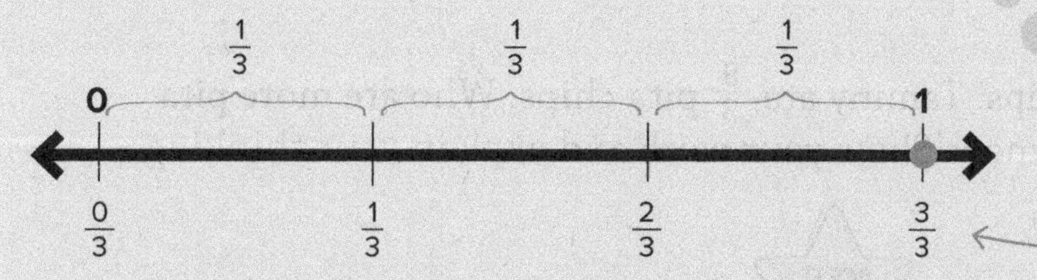

think!
When the top and bottom numbers are the same, the fraction is the same as one whole.

See how the distance between 0 and 1 is broken into 3 equal parts? You can count from 0 to 1 in thirds. When you get to $\frac{3}{3}$, you are at the whole number 1.

You can also have fractions worth more than one. This happens when the top number is greater than the bottom number (when the numerator is greater than the denominator).

Look at the number $\frac{8}{4}$.

The bottom number tells you the number of equal parts in each whole. It takes two wholes to get to 8 fourths. This means $\frac{8}{4}$ = 2 wholes.

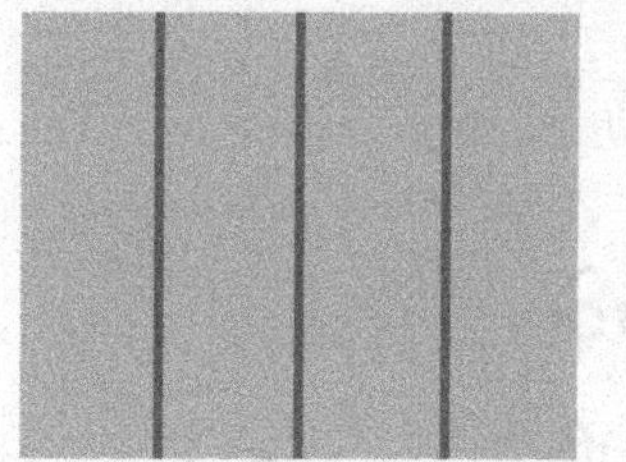

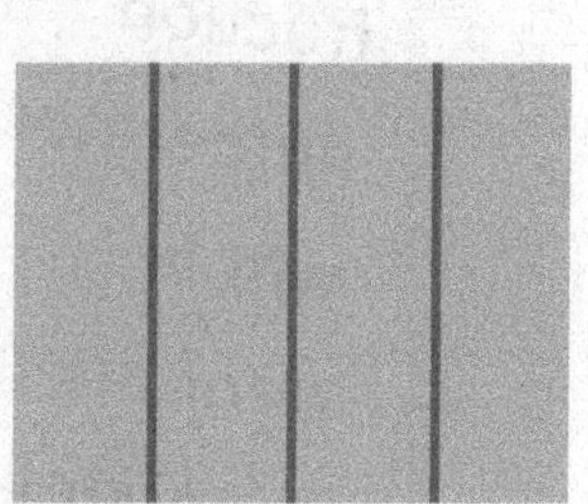

Fractions greater than one on a number line.

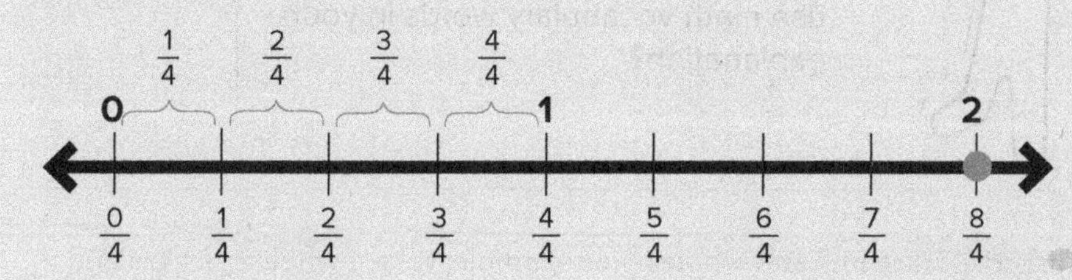

think!
You can use division to find out how many wholes are in a fraction. Divide the top number by the bottom number. 8 ÷ 4 = 2 wholes.

Look at the number line. Notice that $\frac{8}{4}$ is at the same point as the whole number 2. It is clear that $\frac{4}{4}$ is equal to 1 whole, so $\frac{8}{4}$ is equal to 2 wholes.

PRACTICE: Now you try

Write the fraction and whole number for the shaded parts.

1.

Fraction ____________________

Whole Number ____________________

2.

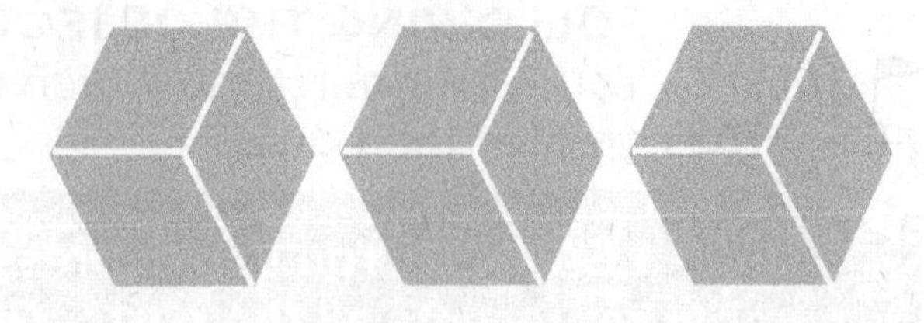

Fraction ____________________

Whole Number ____________________

Lloyd ate 6 whole pita chips. Tammy ate $\frac{9}{3}$ pita chips. Who ate more pita chips? Explain how you know. Show your work and explain your thinking on a piece of paper.

ACE IT TIME!

	yes	no
Did you underline the question in the word problem?		
Did you circle the numbers or number words?		
Did you box the clue words that tell you what operation to use?		
Did you use a picture to show your thinking?		
Did you label your numbers and your picture?		
Did you explain your thinking and use math vocabulary words in your explanation?		

Math Vocabulary

- fraction
- numerator
- whole
- denominator
- number line

Math on the Move

You can use food to practice fractions and wholes. For example, take 4 whole crackers and divide each cracker into fourths. Divide the fourths again and you will have 8 fourths. Continue and you will have 12 fourths and 16 fourths. You can practice this with many different fractions.

Compare Fractions with the Same Denominator

FOLLOWING THE OBJECTIVE
You will compare two fractions that have the same denominator (bottom number).

LEARN IT: You can compare fractions with the same denominator by using a number line or by reasoning.

To compare, place the fractions on a number line. Use greater than (>) and less than (<) symbols.

Compare $\frac{5}{6}$ and $\frac{2}{6}$.

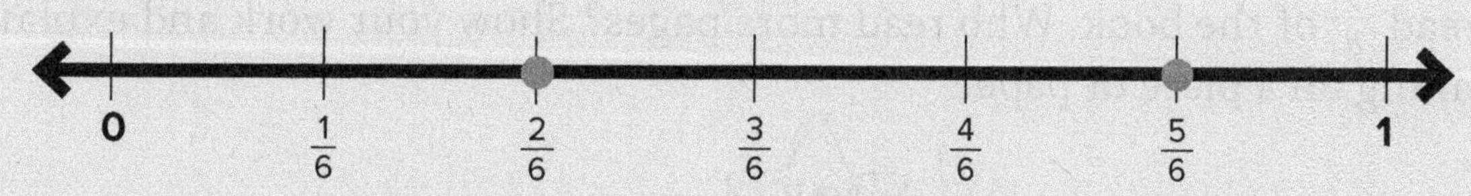

Look at $\frac{2}{6}$ and $\frac{5}{6}$ on the number line. Which is closer to 1?

$\frac{5}{6}$ is closer to 1, so it is larger. $\frac{5}{6} > \frac{2}{6}$

You can also compare by finding the fraction closer to 0.

$\frac{2}{6}$ is closer to 0, so it is smaller. $\frac{2}{6} < \frac{5}{6}$

You can also use reasoning to compare fractions with the same denominator. Use greater than (>) or less than (<) symbols.

Compare $\frac{4}{8}$ and $\frac{6}{8}$.

The bottom numbers are the same. This means the fractions represent the same size parts. The top number tells you how many parts each fraction has. Compare the top numbers to see which is larger and which is smaller.

$4 < 6$ $\quad$ $6 > 4$

$\frac{4}{8} < \frac{6}{8}$ $\quad$ $\frac{6}{8} > \frac{4}{8}$

think!
Would you rather have 2 out of 6 ($\frac{2}{6}$) pieces of your favorite candy bar? Or would you rather have 5 out of 6 ($\frac{5}{6}$) pieces of your favorite candy bar? Why?

PRACTICE: Now you try

Use the number line to compare fractions. Write > or <.

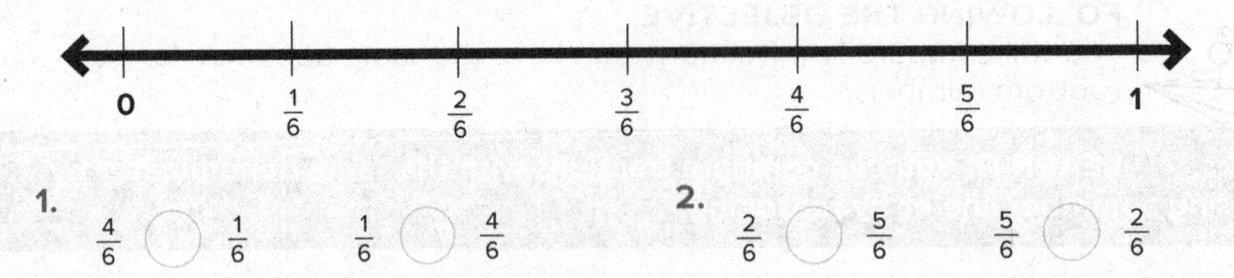

1. $\frac{4}{6}$ ◯ $\frac{1}{6}$ $\frac{1}{6}$ ◯ $\frac{4}{6}$

2. $\frac{2}{6}$ ◯ $\frac{5}{6}$ $\frac{5}{6}$ ◯ $\frac{2}{6}$

Charlene and Pat bought the same book from the book fair. They both started reading at the same time. After two days, Charlene had read $\frac{5}{8}$ of the book and Pat had read $\frac{2}{8}$ of the book. Who read more pages? Show your work and explain your thinking on a piece of paper.

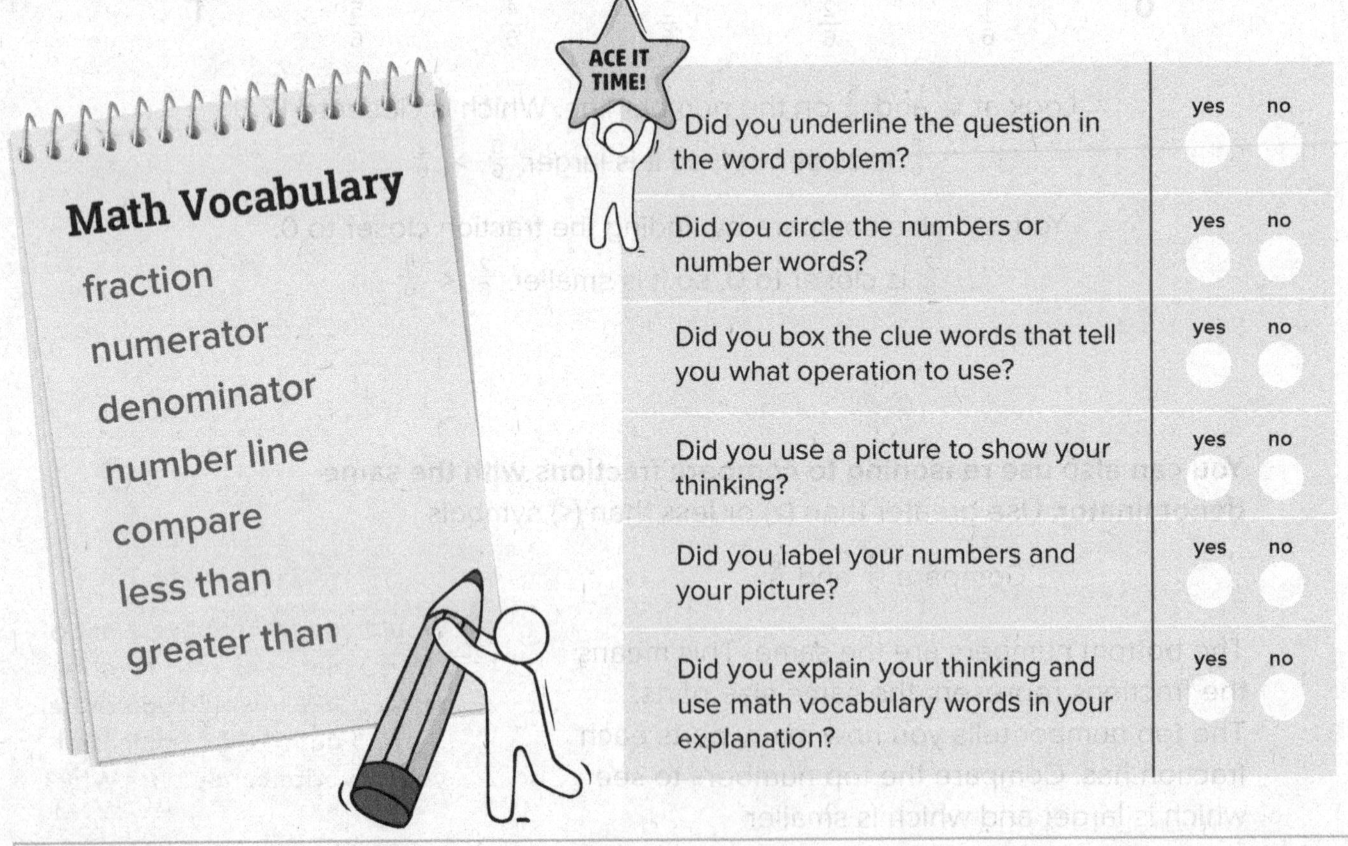

	yes	no
Did you underline the question in the word problem?		
Did you circle the numbers or number words?		
Did you box the clue words that tell you what operation to use?		
Did you use a picture to show your thinking?		
Did you label your numbers and your picture?		
Did you explain your thinking and use math vocabulary words in your explanation?		

You can practice comparing fractions with items around the house. Start by dividing a whole into equal parts. Then compare the parts. Try to understand the concept by answering questions like, "Are 3 parts out of 4 larger than 2 parts out of 4?"

Compare Fractions with the Same Numerator

FOLLOWING THE OBJECTIVE
You will compare two fractions that have the same numerator.

LEARN IT: You can compare fractions with the same numerator by using a number line or by reasoning.

To compare, place the fractions on a number line. Use greater than (>) and less than (<) symbols.

Compare $\frac{3}{4}$ and $\frac{3}{6}$.

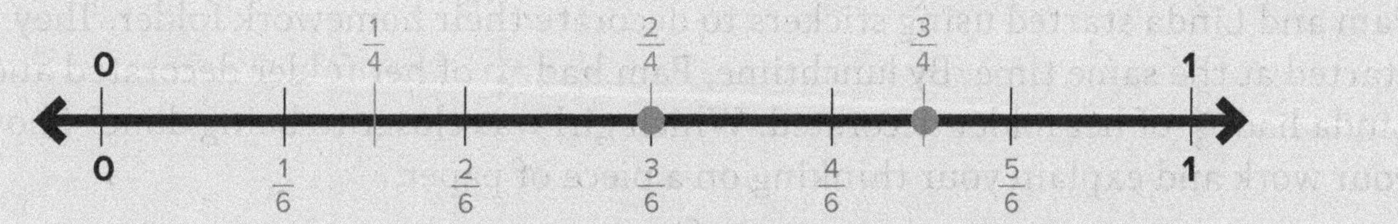

Look at $\frac{3}{4}$ and $\frac{3}{6}$ on the number line. Notice how they have different denominators. This means they represent different parts. You have to divide into four parts for $\frac{3}{4}$. You have to divide into six parts for $\frac{3}{6}$.

Which number is closer to 1? $\frac{3}{4} > \frac{3}{6}$ Which number is closer to 0? $\frac{3}{6} < \frac{3}{4}$

You can also use reasoning to compare fractions with the same denominator.
Use greater than (>) or less than (<) symbols.

Compare $\frac{1}{2}$ and $\frac{1}{3}$.

The top numbers are the same. This means the fractions have the same number of parts. Because the bottom numbers are different, the parts are different sizes.

The bigger the bottom number, the more pieces the whole is broken into. If you have to break a pie into three pieces, you will have smaller pieces than if you only broke it into halves. When the top numbers are the same, the fraction with the **smaller** bottom number is larger.

Think about denominators.

	$2 < 3$		$3 > 2$
So . . .	$\frac{1}{2} > \frac{1}{3}$	and	$\frac{1}{3} < \frac{1}{2}$

think!
Would you want one small piece or one big piece?

PRACTICE: Now you try

Use the number line to compare fractions. Write > or <.

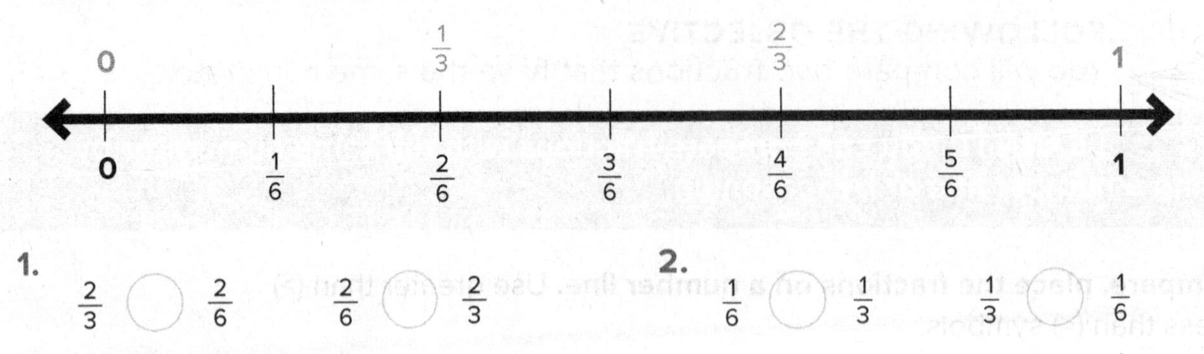

1. $\frac{2}{3}$ ◯ $\frac{2}{6}$ $\frac{2}{6}$ ◯ $\frac{2}{3}$

2. $\frac{1}{6}$ ◯ $\frac{1}{3}$ $\frac{1}{3}$ ◯ $\frac{1}{6}$

Pam and Linda started using stickers to decorate their homework folder. They started at the same time. By lunchtime, Pam had $\frac{3}{4}$ of her folder decorated and Linda had $\frac{3}{8}$ of her folder decorated. Which girl was closer to being done? Show your work and explain your thinking on a piece of paper.

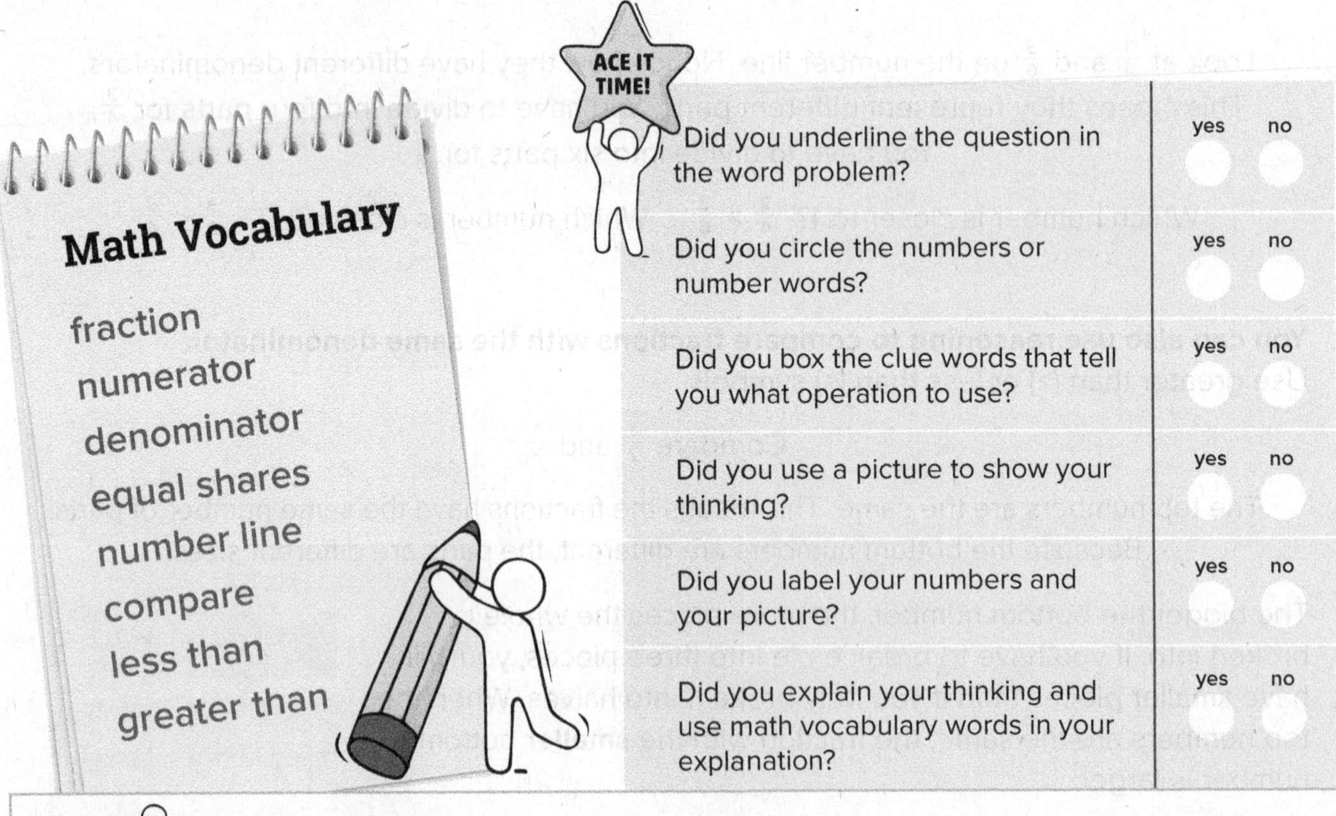

	yes	no
Did you underline the question in the word problem?	◯	◯
Did you circle the numbers or number words?	◯	◯
Did you box the clue words that tell you what operation to use?	◯	◯
Did you use a picture to show your thinking?	◯	◯
Did you label your numbers and your picture?	◯	◯
Did you explain your thinking and use math vocabulary words in your explanation?	◯	◯

Math on the Move

Measuring cups and measuring spoons are great tools for comparing fractions. Which is larger: $\frac{1}{2}$ of a teaspoon or $\frac{1}{4}$ of a teaspoon? Would you rather have $\frac{1}{2}$ cup of orange juice or $\frac{1}{4}$ cup?

Comparing and Ordering Fractions

FOLLOWING THE OBJECTIVE
You will compare and order fractions.

LEARN IT: Remember that the size of a fraction depends on the size of a whole. One-fourth of a small pizza will not be the same size as one-fourth of a large pizza. You can't compare fractions that represent parts of a different whole. But you can compare fractions related to the same whole.

Let's assume the fractions below come from the same whole.

Ordering Fractions with the Same Numerator. Put fractions $\frac{1}{2}$, $\frac{1}{4}$, and $\frac{1}{3}$ in order from least to greatest.

Remember from the previous lessons that the denominator determines the size of each part. The larger the bottom number, the smaller the part. The smaller the bottom number, the larger the part.

Think about denominators.

	$2 < 3$	and	$3 < 4$
So . . .	$\frac{1}{2} > \frac{1}{3}$	and	$\frac{1}{3} > \frac{1}{4}$

$\frac{1}{2}$ is greater than $\frac{1}{3}$ and $\frac{1}{3}$ is greater than $\frac{1}{4}$.

This means $\frac{1}{4}$ is less than $\frac{1}{3}$, and $\frac{1}{3}$ is less than $\frac{1}{2}$.

Ordering Fractions with the Same Denominator. Put fractions $\frac{4}{6}$, $\frac{1}{6}$, and $\frac{5}{6}$ in order from least to greatest.

Remember from previous lessons that the same denominator means the same size of parts. The numerator determines how big the fraction's value is.

Think about numerators.

	$1 < 4$	and	$4 < 5$
So . . .	$\frac{1}{6} < \frac{4}{6}$	and	$\frac{4}{6} < \frac{5}{6}$

$\frac{1}{6}$ is less than $\frac{4}{6}$, and $\frac{4}{6}$ is less than $\frac{5}{6}$.

PRACTICE: Now you try

Write the fractions in order from **least to greatest**.

1. $\frac{1}{3}$, $\frac{1}{2}$, $\frac{1}{8}$, ____, ____, ____

2. $\frac{4}{6}$, $\frac{2}{6}$, $\frac{6}{6}$, ____, ____, ____

3. $\frac{2}{8}$, $\frac{1}{8}$, $\frac{6}{8}$, ____, ____, ____

4. $\frac{2}{4}$, $\frac{3}{4}$, $\frac{1}{4}$, ____, ____, ____

5. $\frac{2}{6}$, $\frac{2}{3}$, $\frac{2}{5}$, ____, ____, ____

6. $\frac{3}{6}$, $\frac{3}{3}$, $\frac{3}{4}$, ____, ____, ____

Andrew decided that he would hike the trail behind his house for three days. On Monday, he walked $\frac{3}{4}$ of a mile. On Tuesday, he walked $\frac{3}{8}$ of a mile. On Wednesday, he walked $\frac{3}{3}$ of a mile. Which day did he walk the most? Show your work and explain your thinking on a piece of paper.

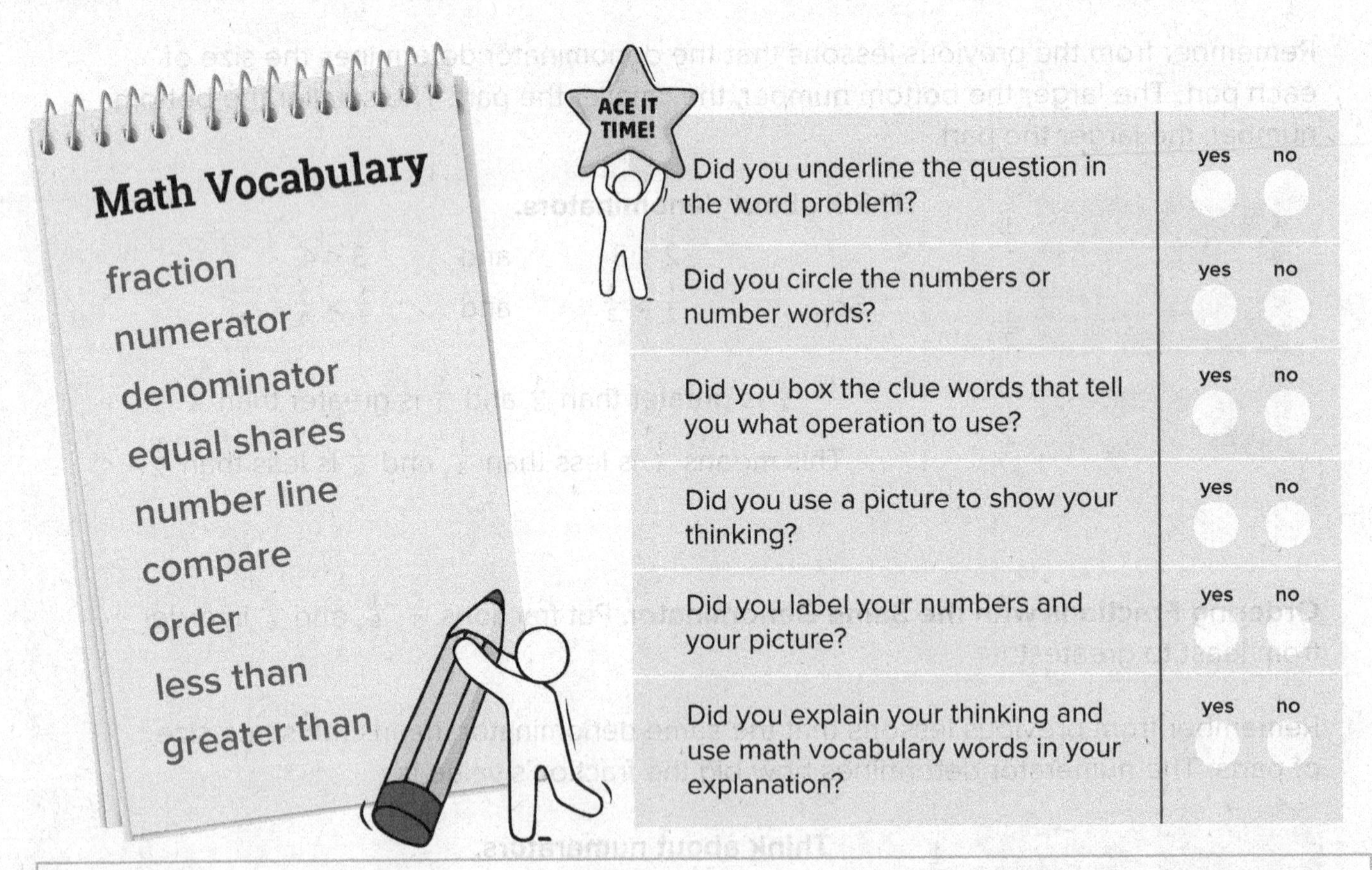

	yes	no
Did you underline the question in the word problem?		
Did you circle the numbers or number words?		
Did you box the clue words that tell you what operation to use?		
Did you use a picture to show your thinking?		
Did you label your numbers and your picture?		
Did you explain your thinking and use math vocabulary words in your explanation?		

Math on the Move

Practice ordering fractions. Ask an adult to write fractions on index cards. Using what you have learned in this lesson, put them in order from least to greatest and from greatest to least.

Equivalent Fractions

FOLLOWING THE OBJECTIVE
You will find and write equivalent fractions.

LEARN IT: Equivalent fractions are fractions that have different names but the same value. They will be at the same point on a number line.

Use a number line to find the equivalent fraction. Turkey slices are on sale at the local supermarket. The slices come in packages of $\frac{1}{6}$ pound. Lisa needs $\frac{2}{3}$ pound. How many $\frac{1}{6}$ bags does she need to buy?

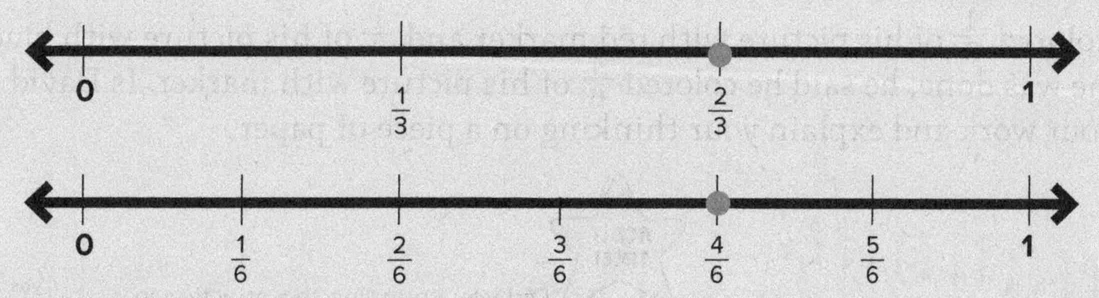

Look at the number lines. See how the zeros and ones line up? This means the whole is the same size. One number line is divided into thirds. The other number line is divided into sixths. What other numbers line up? This means they are also the same size.

Do you see how $\frac{1}{3}$ and $\frac{2}{6}$ are the same? And also how $\frac{2}{3}$ and $\frac{4}{6}$ are the same?

Lisa needs $\frac{2}{3}$ pounds of turkey. This is the same as $\frac{4}{6}$ pounds of turkey. She should buy four $\frac{1}{6}$ pound bags.

You can also use models to show equivalent fractions. Diane cut a turkey sub into 3 equal pieces. She is going to share it with her two sisters. They will each get one piece. How many pieces would the sisters get if Diane decided to cut the sub into 6 pieces instead?

See how the subs are drawn the same size? They are cut into 3 equal pieces and 6 equal pieces. Which numbers are the same?

Each sister can have $\frac{1}{3}$ or $\frac{2}{6}$ of a sandwich.

PRACTICE: Now you try

1. Fill in the missing fractions on the number line. Draw a circle around the equivalent fractions.

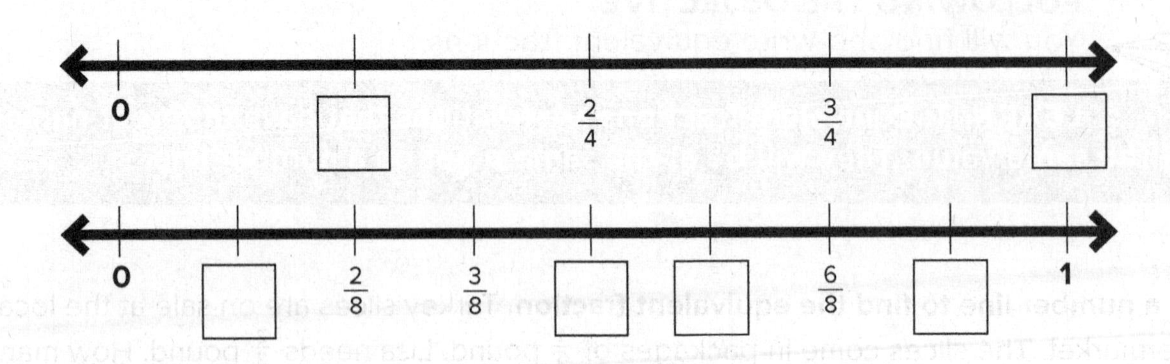

David colored $\frac{1}{3}$ of his picture with red marker and $\frac{1}{3}$ of his picture with blue marker. When he was done, he said he colored $\frac{4}{6}$ of his picture with marker. Is David right? Show your work and explain your thinking on a piece of paper.

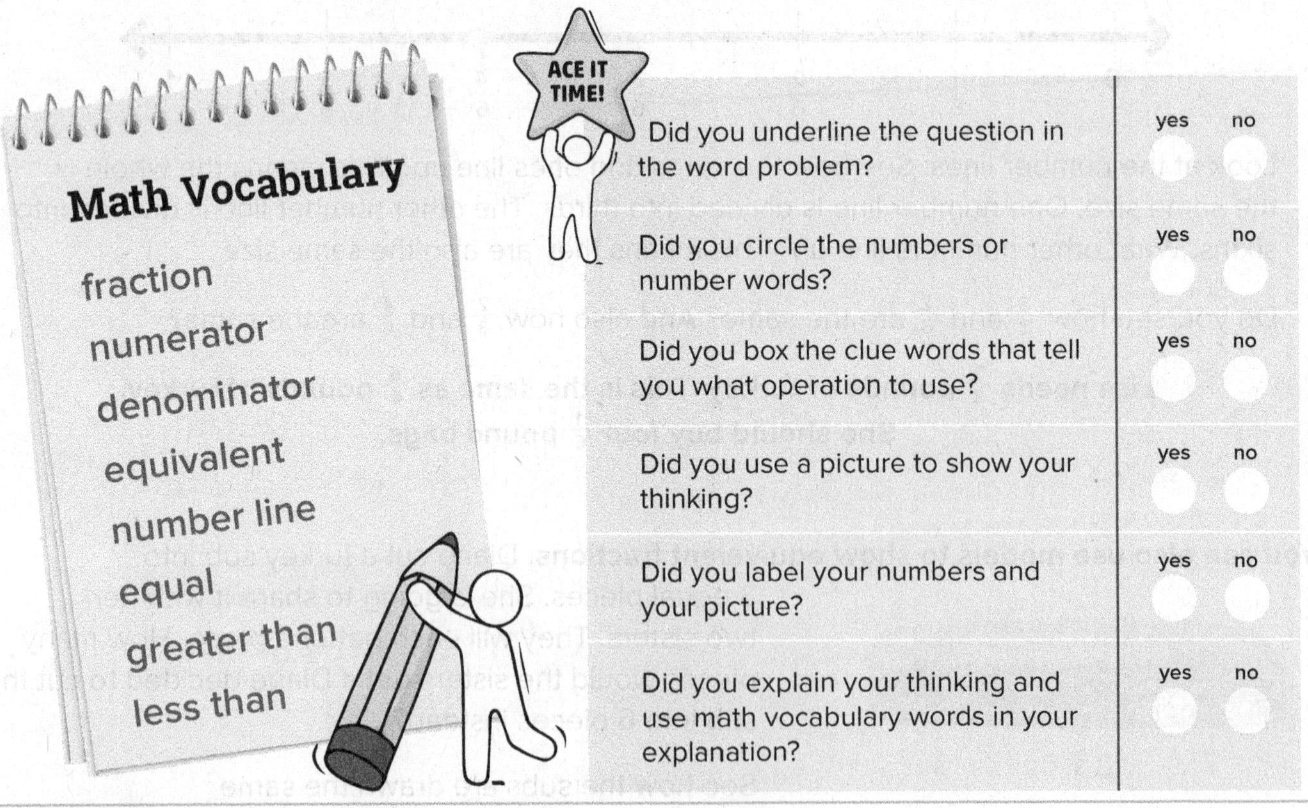

	yes	no
Did you underline the question in the word problem?		
Did you circle the numbers or number words?		
Did you box the clue words that tell you what operation to use?		
Did you use a picture to show your thinking?		
Did you label your numbers and your picture?		
Did you explain your thinking and use math vocabulary words in your explanation?		

Math on the Move

You can practice finding equivalent fractions using measuring cups and measuring spoons. How many $\frac{1}{2}$ cups does it take to fill 1 cup? How many $\frac{1}{4}$ cups fill $\frac{1}{2}$ cup?

REVIEW

Stop and think about what you have learned.

Congratulations! You've finished the lessons for this unit. This means you've learned about fractions and their equal parts. You've learned about unit fractions. You've practiced using a number line to help you solve fraction problems. You can even recognize, create, and explain equivalent fractions.

Now it's time to prove your skills with fractions. Solve the problems below! Use all of the methods you have learned.

Activity Section

Draw lines to make equal shares. Write the fraction.

1. Sheila and her three friends share three rice cakes equally. ______

2. Three siblings share four oatmeal bars equally. ______

Name and write the fraction for the shaded parts.

3.

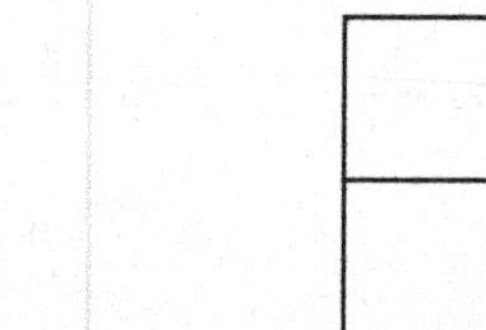

4.

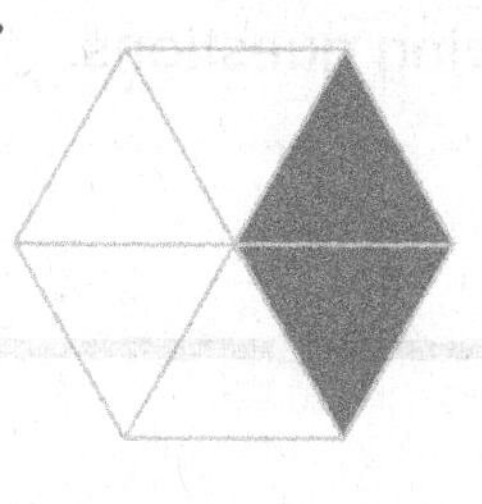

5.

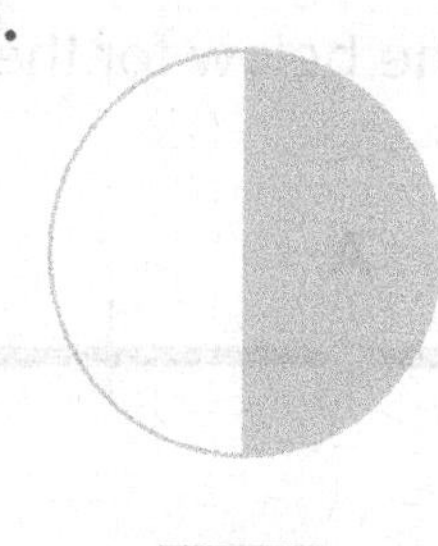

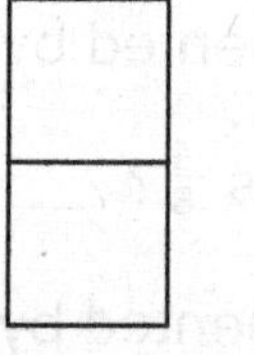

6.

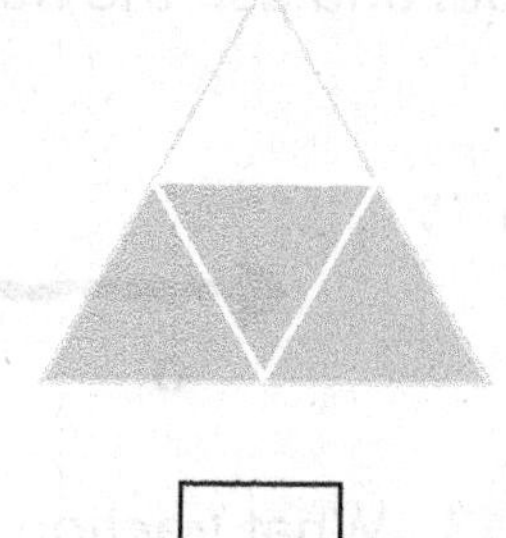

Shade the models to represent the fractions.

7.	8.	9.	10.
	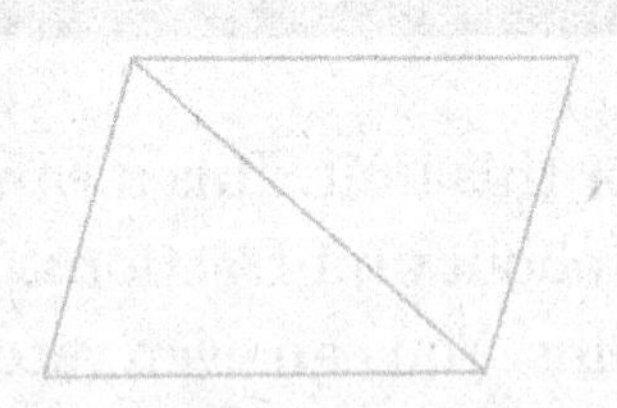		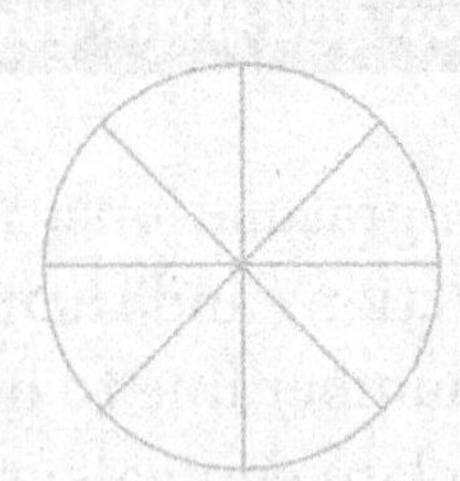
$\frac{2}{3}$	$\frac{1}{2}$	$\frac{1}{4}$	$\frac{3}{8}$

Fill in the missing fractions on the number lines.

11.

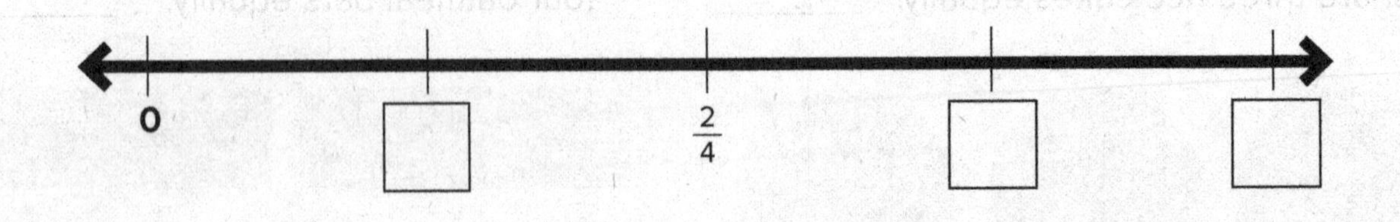

12.

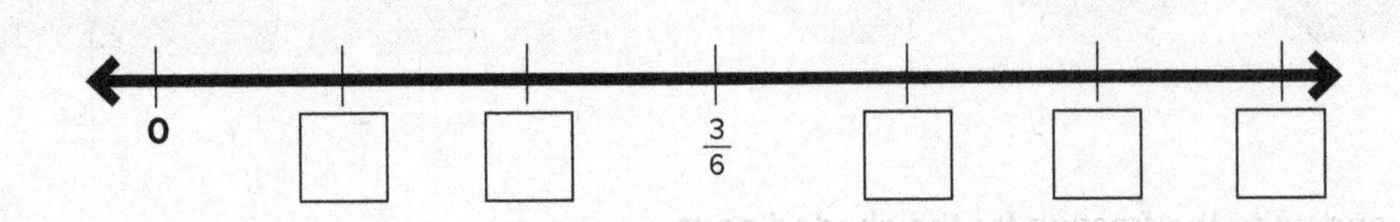

Label and use the number line below for the following questions.

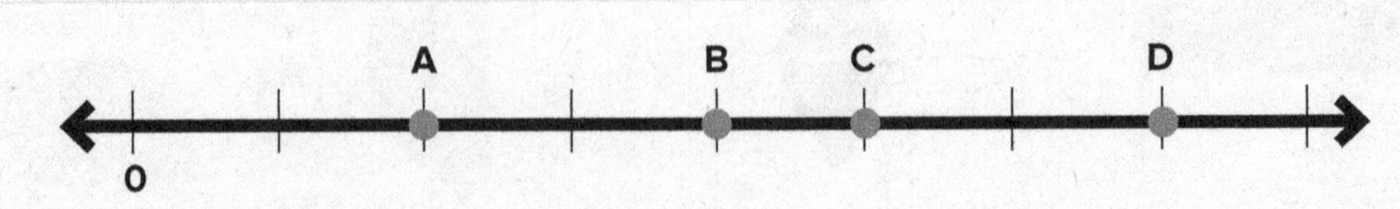

13. What fraction is represented by point D? ________

14. Which point represents $\frac{4}{8}$? ________

15. What fraction is represented by point A? ________

Fill in the missing fractions on the number lines.

16. $\frac{3}{3}$ = ______	**17.** 4 = ______	**18.** $\frac{8}{4}$ = ______
19. 3 = ______	**20.** 1 = ______	**21.** $\frac{6}{2}$ = ______

Compare fractions. Write $>$ or $<$.

22. $\frac{3}{8}$ ◯ $\frac{3}{6}$	**23.** $\frac{2}{4}$ ◯ $\frac{1}{4}$	**24.** $\frac{1}{3}$ ◯ $\frac{3}{3}$	**25.** $\frac{2}{4}$ ◯ $\frac{2}{6}$
26. $\frac{5}{8}$ ◯ $\frac{2}{8}$	**27.** $\frac{1}{6}$ ◯ $\frac{1}{3}$	**28.** $\frac{5}{6}$ ◯ $\frac{2}{6}$	**29.** $\frac{4}{4}$ ◯ $\frac{4}{6}$
30. $\frac{1}{2}$ ◯ $\frac{2}{2}$	**31.** $\frac{5}{6}$ ◯ $\frac{5}{8}$	**32.** $\frac{2}{6}$ ◯ $\frac{2}{3}$	**33.** $\frac{2}{4}$ ◯ $\frac{3}{4}$

Write the fractions in order from **least to greatest**.

34. $\frac{1}{4}, \frac{1}{2}, \frac{1}{3}$ ______, ______, ______

35. $\frac{6}{8}, \frac{2}{8}, \frac{5}{8}$ ______, ______, ______

36. $\frac{2}{6}, \frac{1}{6}, \frac{6}{6}$ ______, ______, ______

Write the fractions in order from **greatest to least**.

37. $\frac{1}{8}, \frac{1}{2}, \frac{1}{6}$ ______, ______, ______

38. $\frac{1}{8}, \frac{7}{8}, \frac{3}{8}$ ______, ______, ______

39. $\frac{3}{4}, \frac{1}{4}, \frac{4}{4}$ ______, ______, ______

Draw lines in models and shade to show equivalent fractions. Fill out missing fractions.

40. $\frac{2}{4} = \frac{\square}{8}$

41. $\frac{2}{3} = \frac{\square}{6}$

42.  $\frac{4}{4} = \frac{\square}{3}$

Understand the meaning of what you have learned and apply your knowledge.

You can recognize, write, and explain equivalent fractions.

Activity Section

Tiffany and Lisa bought two cheesy flatbreads that were exactly the same size. Tiffany divided her flatbread into 4 equal pieces, and Lisa divided her flatbread into 8 equal pieces. Tiffany ate 3 of her 4 pieces and Lisa ate 6 of her 8 pieces. Lisa says that they ate the same amount. Is she right? Explain your answer and show your work.

DISCOVER

Discover how you can apply the information you have learned.

You can compare two fractions with the same numerator or the same denominator by reasoning.

Activity Section

The children in Mrs. Cummings's class were having an argument about who had the longest pencil. They decided to settle the argument by measuring the length of their pencils using a ruler. Melissa's pencil was 9 inches, Lisa's was $\frac{2}{4}$ of a foot, Terry's was $\frac{4}{6}$ of a foot, and Brian's was $\frac{1}{3}$ of a foot. Who had the longest pencil? Show your work and explain your answer. *Hint:* Draw number lines for each pencil and compare.

Data Concepts

Picture Graphs and Bar Graphs

FOLLOWING THE OBJECTIVE
You will understand and create picture and bar graphs.

LEARN IT: Picture and bar graphs are visual ways to represent data. Notice how the same information can be shown as both a picture and a bar graph.

Picture Graph
Use pictures or symbols to show the value of the data.

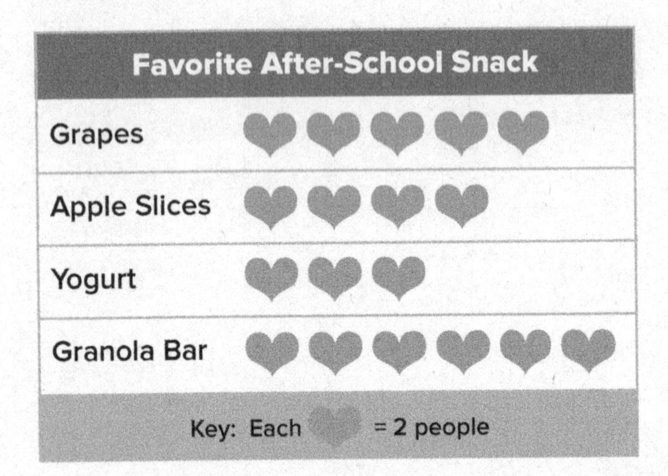

Bar Graph
Use rectangular bars to show the value of the data.

Look at each graph and determine how many people enjoy each of these after-school snacks. In the picture graph, we understand from the key that each heart represents two people. For example: grapes have five hearts, which we can calculate as 2 + 2 + 2 + 2 + 2 = 10 or use multiplication, 5 × 2 = 10. In reading the bar graph, the top of the bar is in line with the number 10, which indicates that 10 people enjoy grapes as their favorite after-school snack.

We can use either of the graphs above to answer these questions.

1. How many more people like granola bars than apple slices?

   12 – 8 = 4
   (granola) – (apples) = 4 more

2. How many total people like granola bars and grapes?

   12 + 10 = 22
   (granola) + (grapes) = 22 total

PRACTICE: Now you try

1. The third-grade students at Meadow Brook Elementary voted for their favorite subjects in school. The results are in the chart below.

   a. Use the data from the chart "Favorite Subject in School" to create both a picture and a bar graph below.

   b. For the picture graph, create a symbol for the key and determine how many students it stands for.

   c. Complete each graph.

Favorite Subject in School	
Reading	12 students
Math	14 students
Writing	8 students
Science	10 students

Picture Graph

Draw the correct number of pictures for each subject choice.

Favorite Subject In School	
Reading	
Math	
Writing	
Science	
Key: Each ______ = ______ students	

Bar Graph

Draw and shade a bar to show the number for each subject choice.

Favorite Subject In School

14
12
10
8
6
4
2
0

Reading Math Writing Science

School Subjects

After creating the graphs, use the information to answer each of these questions.

2. How many total students are in the third grade at Meadow Brook Elementary?

3. How many more students enjoy Math than enjoy Writing?

4. How many total students enjoy Math and Science?

5. How many more students enjoy Reading than enjoy Writing?

Using the graphs entitled "Favorite Subject in School," answer this question:
How many more students enjoy Reading or Math than enjoy Science or Writing?

Math Vocabulary

sum
total
difference
subtract
compare

ACE IT TIME!

Did you underline the question in the word problem?	yes	no
Did you circle the numbers or number words?	yes	no
Did you box the clue words that tell you what operation to use?	yes	no
Did you use a picture to show your thinking?	yes	no
Did you label your numbers and your picture?	yes	no
Did you explain your thinking and use math vocabulary words in your explanation?	yes	no

Math on the Move

Think of a survey question to ask family members and/or friends. Collect their responses and create a picture graph or bar graph to represent their responses.

Line Plots

FOLLOWING THE OBJECTIVE
You will understand and create line plot graphs.

LEARN IT: A line plot graph is another way to visually represent information. A line plot records the number of times (frequency) data items occur. Because a line plot represents the "shape" of the data, they are helpful to see groupings or gaps in information.

Line Plots: Use marks, like an "X," to represent data above a number line.

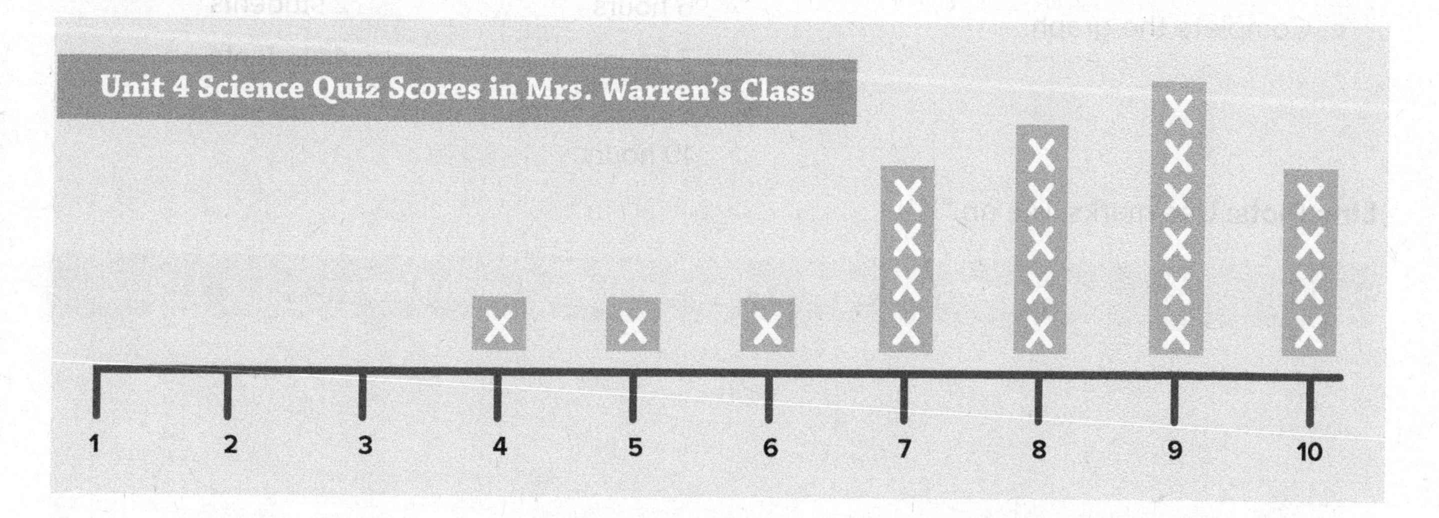

Look at the line plot above and analyze how the students in Mrs. Warren's class did on the Unit 4 science quiz. Since each "X" represents one student, there were four students who answered 7 out of 10 questions correctly. There were also four students who received a perfect score and answered 10 out of 10 questions correctly. Since there are no "X's" on the scores of 1, 2, or 3, no students received those scores.

We use the line plot of Mrs. Warren's science quiz scores above to answer these questions:

1. How many students passed the science quiz with a score of 7 or above?

   4 (score of 7) + 5 (score of 8) + 6 (score of 9) + 4 (score of 10) = 19 students

2. How many more students received a score of 9 than received a score of 8?

   6 (score of 9) – 5 (score of 8) = 1 student

PRACTICE: Now you try

1. The third-grade students in Mrs. Warren's class were required to read at home every night for 20 minutes. Mrs. Warren wanted to show them how much time they had spent reading by making a line plot of their time for one month. Create a line plot below to represent the data in the table.

   a. Use the data in "Hours Spent Reading" to create a line plot graph below.

   b. Create a symbol or use an "X" to represent data.

   c. Complete the graph.

Hours Spent Reading

1 hour, 2 hours	0 students
3 hours	2 students
4 hours	3 students
5 hours	4 students
6 hours	2 students
7 hours	4 students
8 hours, 9 hours, 10 hours	1 student each

Line Plots: Use marks, like an "X"

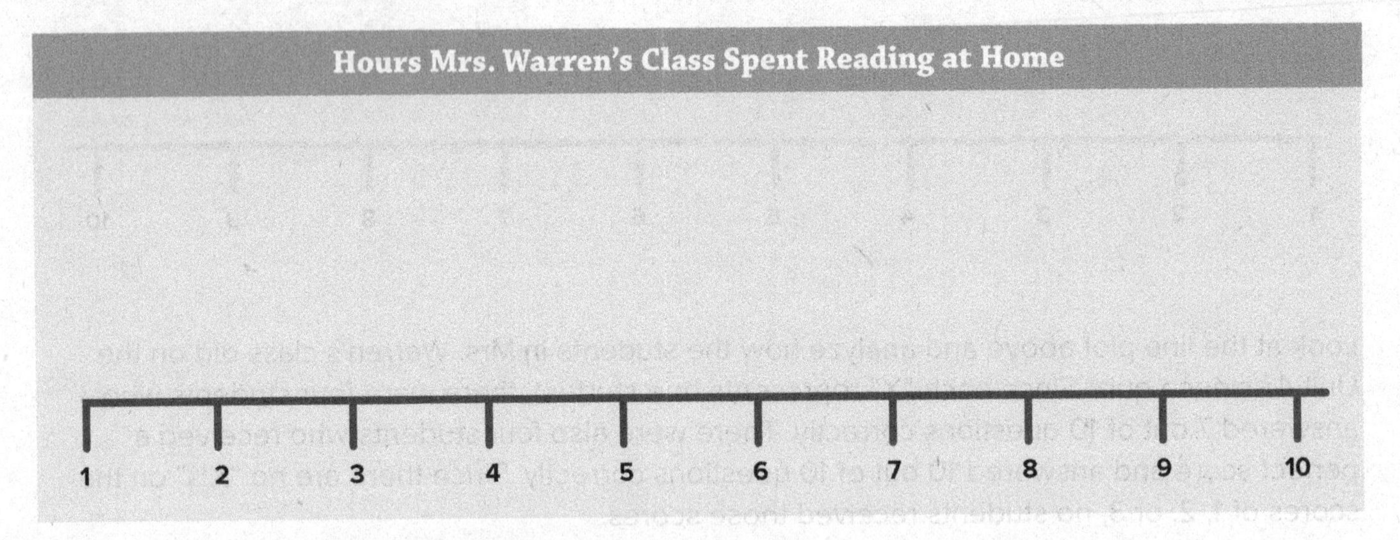

After creating the line plot, use the information to answer each of these questions:

2. How many total students are in Mrs. Warren's third-grade class?

3. How many more students read five hours at home than read three hours?

4. How many students read six or more hours at home?

5. How many students read nine or more hours at home?

Using the graph entitled "Hours Spent Reading at Home in Mrs. Warren's Class," answer this question: how many more students read seven or more hours than read four hours or less? Explain and show your work.

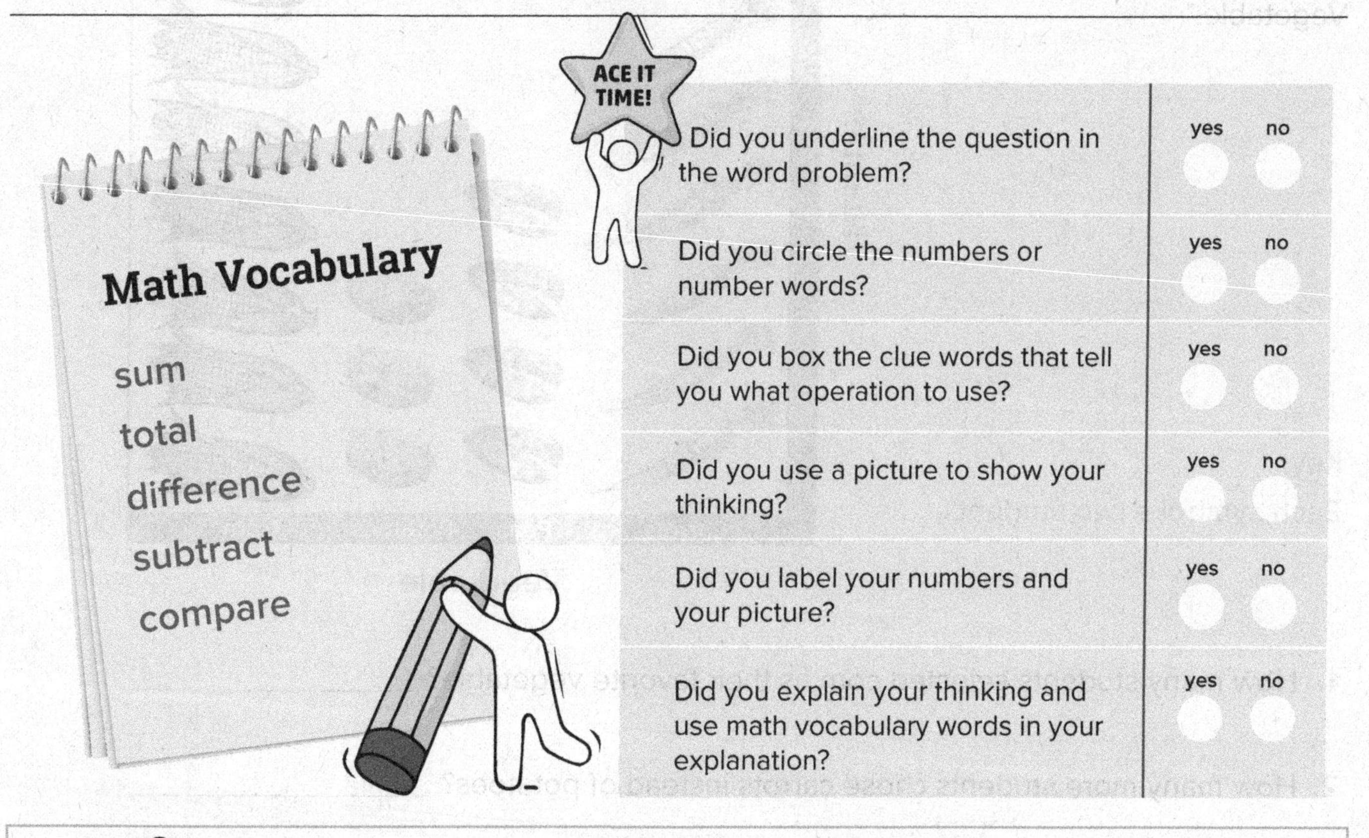

	yes	no
Did you underline the question in the word problem?		
Did you circle the numbers or number words?		
Did you box the clue words that tell you what operation to use?		
Did you use a picture to show your thinking?		
Did you label your numbers and your picture?		
Did you explain your thinking and use math vocabulary words in your explanation?		

Math on the Move

Create a line plot. Count the number of letters in each family member's name and plot them. Make up questions for your family members to answer.

REVIEW

Stop and think about what you have learned.

Congratulations! You've finished the lessons for this unit. This means you've learned how to make and solve problems using picture graphs, bar graphs, and line plots. You understand how to read all three types of graphs, and you can design your own questions and graphs.

Now it's time to prove your skills with graphing and data concepts. Solve the problems below. Use all the methods you have learned.

Activity Section

Interpret Graphs

Answer the questions for the picture graph "Favorite Vegetable."

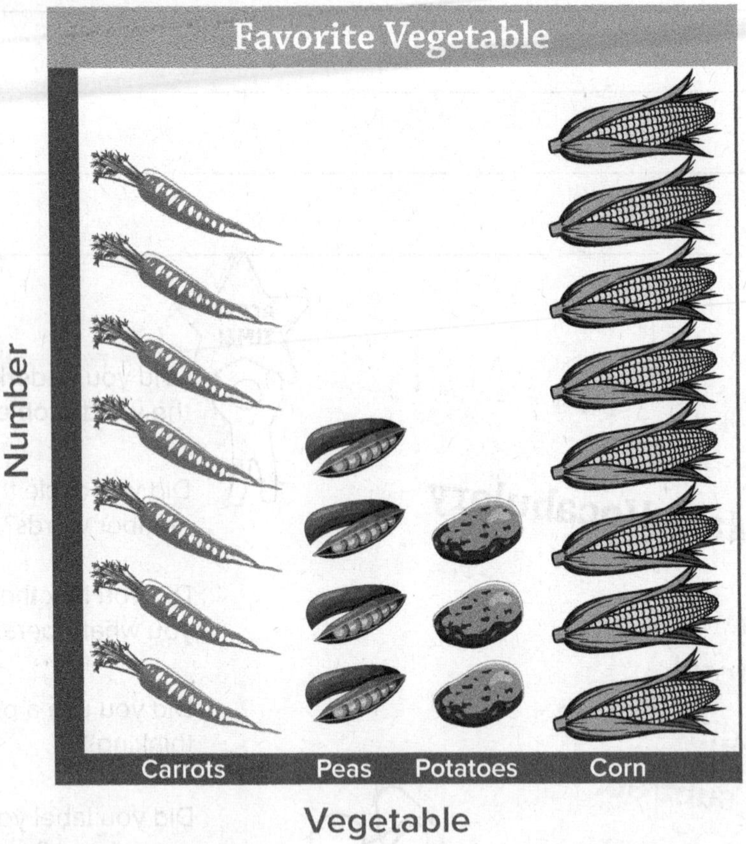

Favorite Vegetable

Number

Carrots Peas Potatoes Corn

Vegetable

Key:
Each symbol = two students

1. How many students selected corn as their favorite vegetable? ____________________

2. How many more students chose carrots instead of potatoes? ____________________

3. How many total students chose root vegetables like carrots and potatoes as their favorite vegetable? ____________________

Answer the questions for the line plots below:

4. How many students read seven or more books over summer break? ___________

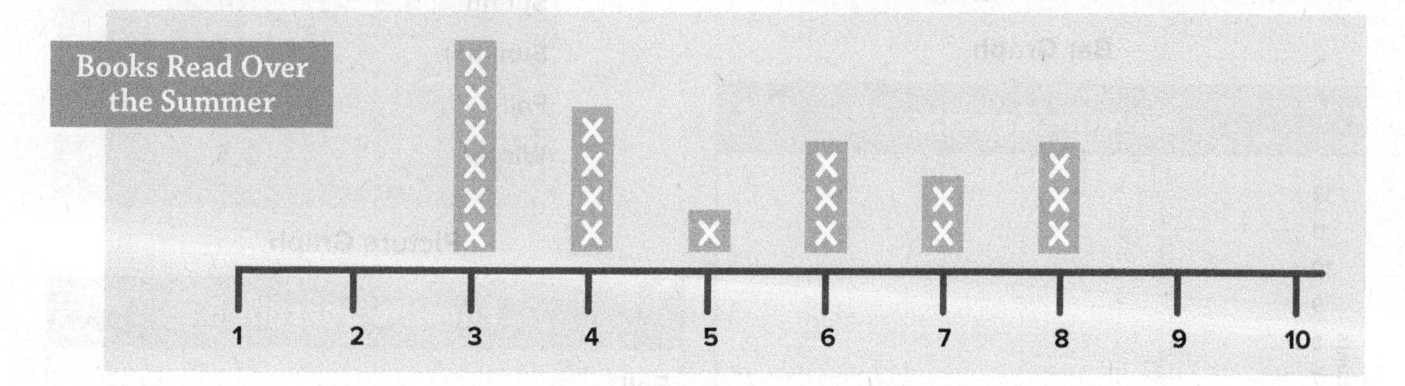

5. How many students received a perfect score on their math quiz? ___________

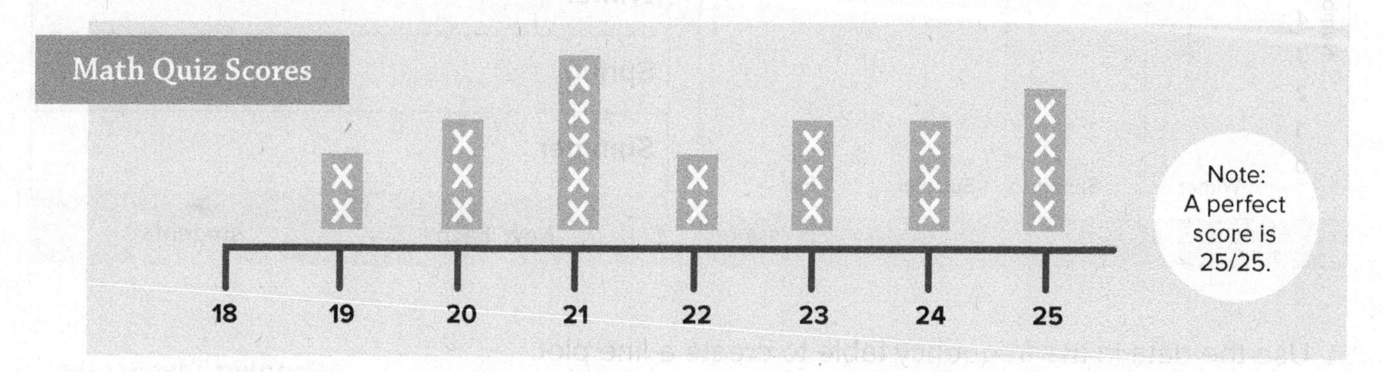

Answer the questions for the bar graph "Number of Push-ups Done in PE."

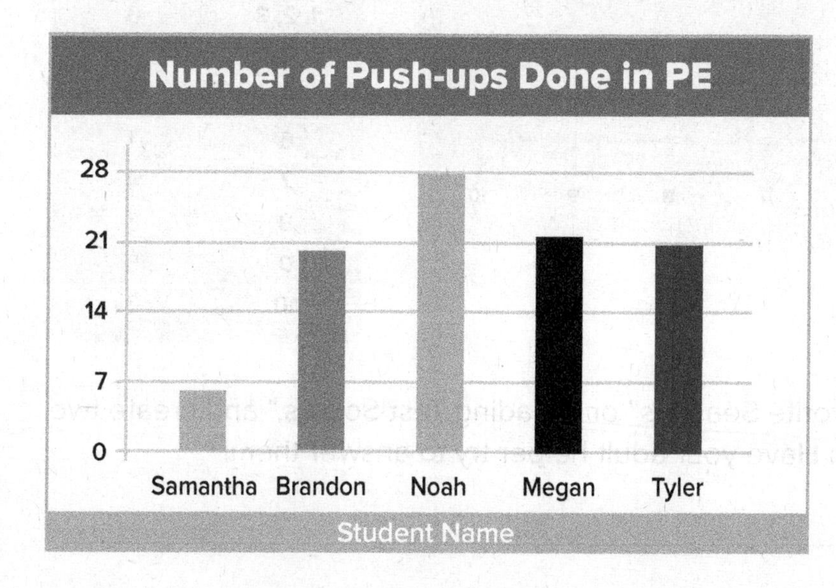

6. How many more push-ups did Noah do than Tyler? ___________

7. Put the students in order from the least number of push-ups to the greatest number of push-ups.

____

____

____

____

____

Create Graphs

8. Use the data in the frequency table to create a bar graph and a picture graph.

Favorite Seasons

Season	Number of Students
Spring	4
Summer	10
Fall	12
Winter	8

Bar Graph

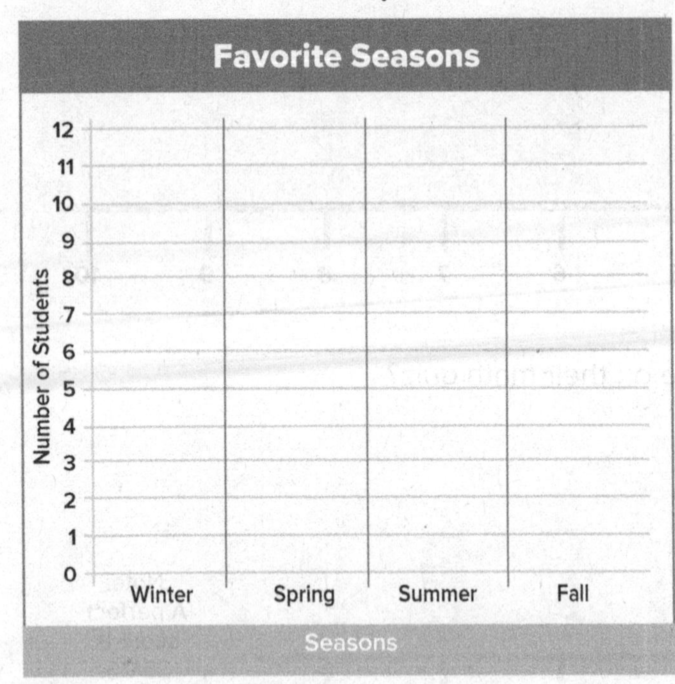

Picture Graph

Favorite Seasons	
Fall	
Winter	
Spring	
Summer	

Key: Each _____ = _____ students

9. Use the data in the frequency table to create a line plot.

Reading Test Scores

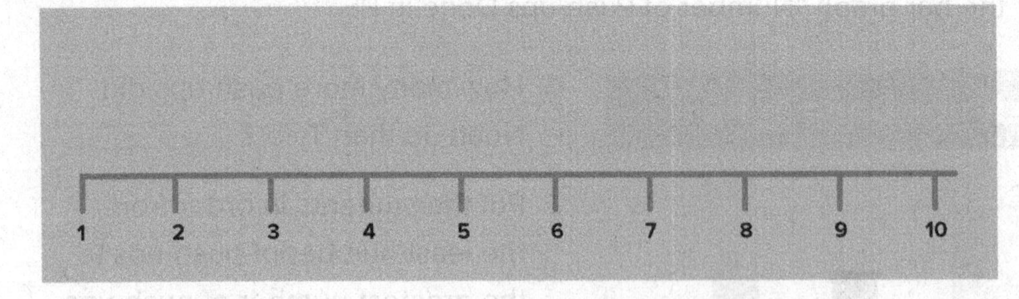

Reading Test Scores

Score	Number of Students
1, 2, 3	0
4	1
5	2
6	1
7	4
8	6
9	5
10	3

Create Questions

Select a graph from above, either "Favorite Seasons" or "Reading Test Scores," and create two questions about the data in that graph. Have your adult helper try to answer them!

10. ____?

11. ____?

UNDERSTAND

Understand the meaning of what you have learned and apply your knowledge.

You can create and interpret bar graphs.

Activity Section

You may only see graphing problems presented in vertical (or up and down) formats. However, picture graphs and bar graphs may also be represented in horizontal (or on its side) formats. Sometimes, this can be confusing. During this activity, you are going to represent data in both vertical and horizontal formats in order to clarify these misunderstandings.

Vertical Format

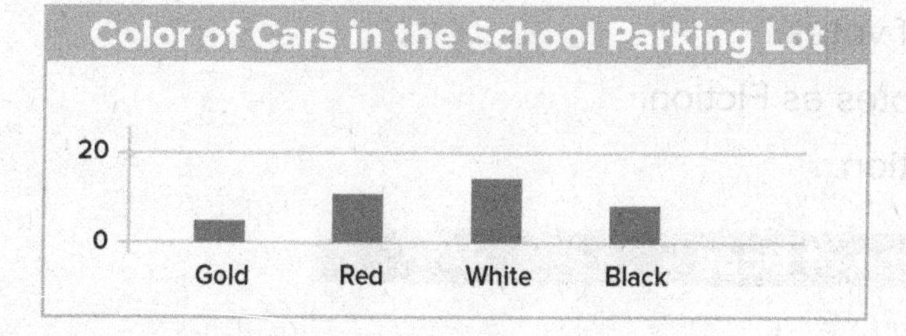

Horizontal Format

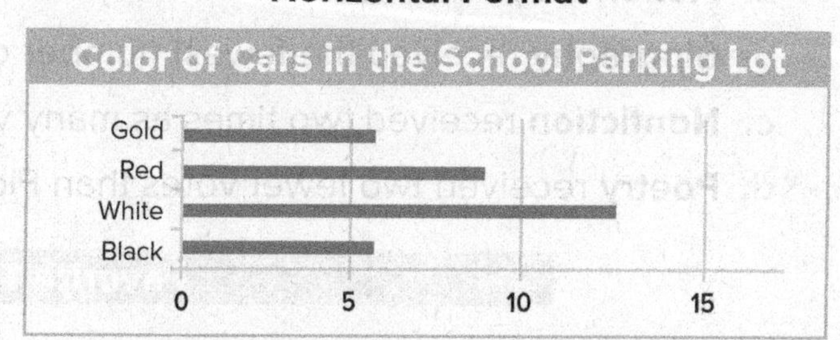

Use the frequency table to create bar graphs in both the vertical and horizontal formats.

Favorite Type of Fiction

Type of Fiction	Number of Students
Historical	2
Fantasy	10
Realistic	8
Science	4

Vertical Format

Horizontal Format

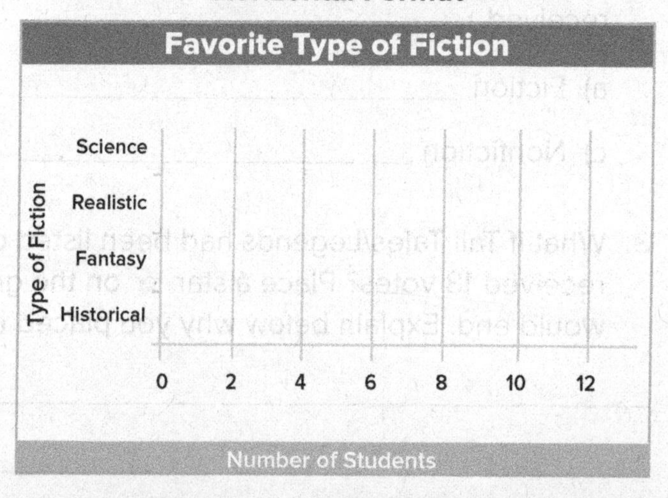

Discover how you can apply the information you have learned.

You can create and interpret picture graphs and bar graphs.
You can add and subtract within 1,000.
You can use multiplication and/or division within 100.

Activity Section

1. Ms. White challenged her students. She created this bar graph but did not list the types of books on the graph. She told her students their task was to place the type of book on its correct bar. She gave them these clues:

   a. **Fiction** received two times as many votes as Biography.

   b. **Biography** received the fewest number of votes.

   c. **Nonfiction** received two times as many votes as Fiction.

   d. **Poetry** received two fewer votes than Fiction.

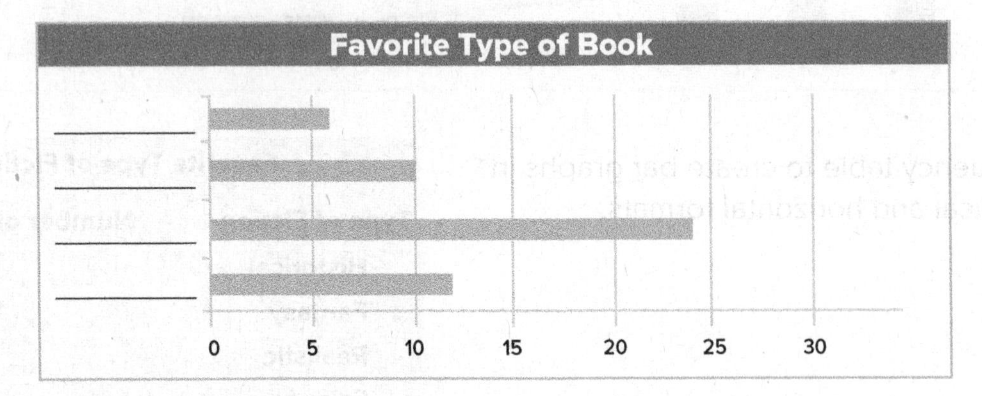

2. Using math sentences, show how you determined the number of votes each book received.

   a) Fiction ______________________ b) Poetry ______________________

   c) Nonfiction ______________________ d) Biography ______________________

3. What if Tall Tales/Legends had been listed on this survey as a favorite genre type and had received 18 votes? Place a star ★ on the graph where the bar you would draw or shade would end. Explain below why you placed the star there.

____

____

Measurement Concepts

Telling and Writing Time

FOLLOWING THE OBJECTIVE
You will learn to tell and write time to the nearest minute.

LEARN IT: You can use an analog clock to tell time.

Telling Time to the Minute

There are 60 minutes in one hour. It takes one hour, or 60 minutes, for the short hand (hour hand) to move from one number to the next. The numbers on the clock represent hours. You might see the short hand between two numbers. The short hand on this clock is pointed between 1 and 2. When writing this time the hour would be 1 o'clock because the short hand has not reached the 2 yet.

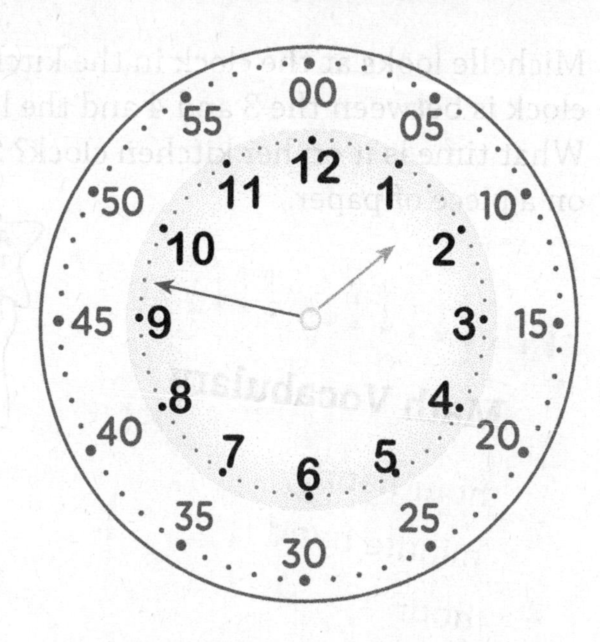

1 : ____
hour : minutes

It takes one minute for the long hand (minute hand) to move from one tick mark on the clock to the next. It takes five minutes for the long hand to move from one number on the clock to another. So it is easy to count by 5s using the numbers to help you. The long hand on this clock is between the 9 and 10. Start at the number 12 and move to the number 1 and count by 5s. 5, 10, 15, 20, 25, 30, 35, 40, 45. Do not count on to the 10 (50), because the long hand is not at the ten yet. When you get to 45, start counting by the minute. Starting at the 9 and counting on by the minute . . . 45, 46, 47.

1 : 47
hour : minutes

PRACTICE: Now you try

Write the time shown on each clock.

Michelle looks at the clock in the kitchen. She notices that the short hand on her clock is between the 3 and 4 and the long hand is 1 minute past the number 8. What time is it on her kitchen clock? Show your work and explain your thinking on a piece of paper.

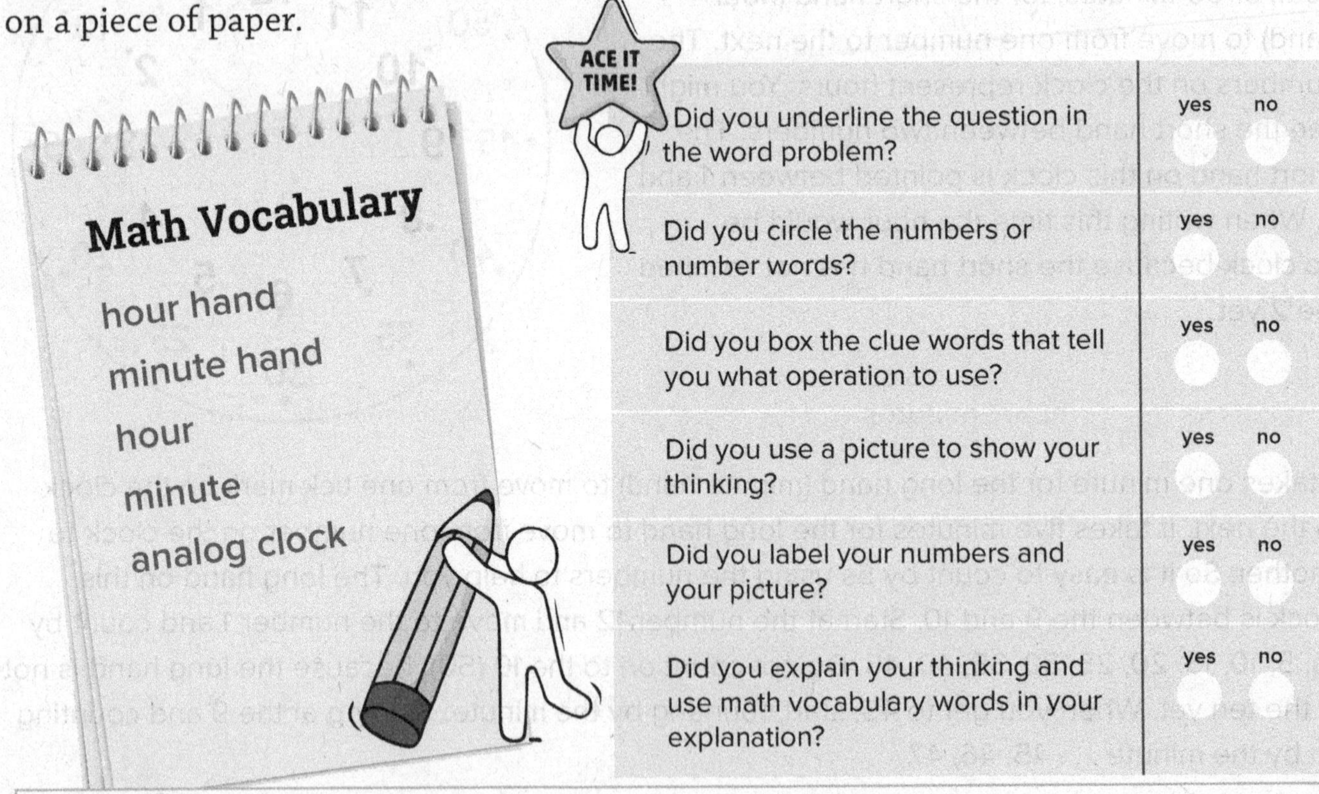

Did you underline the question in the word problem?	yes no
Did you circle the numbers or number words?	yes no
Did you box the clue words that tell you what operation to use?	yes no
Did you use a picture to show your thinking?	yes no
Did you label your numbers and your picture?	yes no
Did you explain your thinking and use math vocabulary words in your explanation?	yes no

Math on the Move

Practice telling and writing time from an analog clock at home. If you do not have an analog clock in the home, simply draw one and label it. You can then use it to practice telling and writing time.

Intervals of Time

FOLLOWING THE OBJECTIVE
You will solve word problems involving addition and subtraction of time intervals in minutes.

LEARN IT: When solving problems involving time, it is important to consider start time, end time, and elapsed time.

Finding the End Time

When solving time interval problems, you may be asked to find the end time. In this case, you will be given a start time and the elapsed time.

Example: Mya leaves to go to the mall at 10:05. It takes her 25 minutes to get there. What time does she get to the mall?

Start Time = 10:05 a.m. **Elapsed Time = 25 minutes** **End Time = ______ : ______**

You can use a **clock** to figure out the time that Mya gets to the store.

Find the starting time on the clock. Skip-count around the clock by 5s and then by 1s until you count the total number of minutes.

Mya gets to the mall at 10:30 a.m.

think!
Start at 10:05. Skip count by 5s for 25 minutes. You end at 10:30 a.m.!

Finding Elapsed Time

When solving time interval problems, you may be asked to find the elapsed time. In this case, you will be given a start time and an end time. **Elapsed time** is the difference between a start time and an end time.

Example: Eric starts his homework at 3:30 p.m. when he comes home from school. He finishes at 4:20 p.m. How long does Eric spend on homework?

Start Time = 3:30 p.m. **Elapsed Time = ______ minutes** **End Time = 4:20 p.m.**

Use a **number line** to figure out how long Eric spent on his homework.

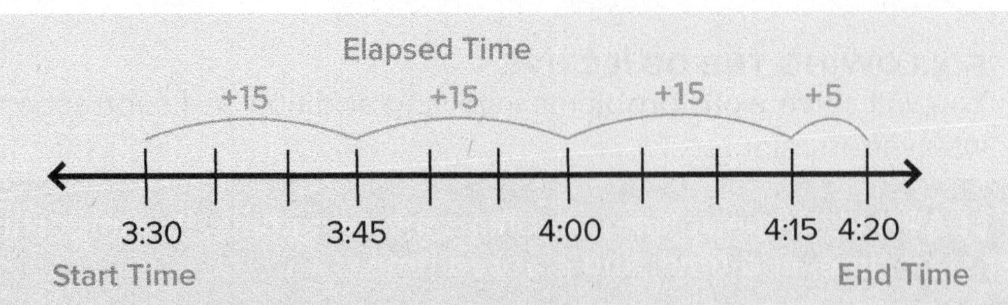

Use a number line to keep track of elapsed time. Start counting in 15 minute intervals to make it easier. Add up the intervals to find the elapsed time.

It took Eric 50 minutes to do his homework.

Finding Start Time

When solving time interval problems, you may be asked to find the start time. In this case, you will be given an end time and an elapsed time.

Example: Julius has been at track practice for 1 hour and 15 minutes. It is now 6:25. What time did he get to practice?

Start Time = _______ : _______ Elapsed Time = 1 hour 15 minutes End Time = 6:25 p.m.

Let's use a **number line** to figure out the time that Julius got to practice. Notice we are starting with the end time and subtracting intervals until they add up to 1 hour and 15 minutes. Subtract in big chunks of time when you can. For example, subtract the hour first, and then go back 15 minutes.

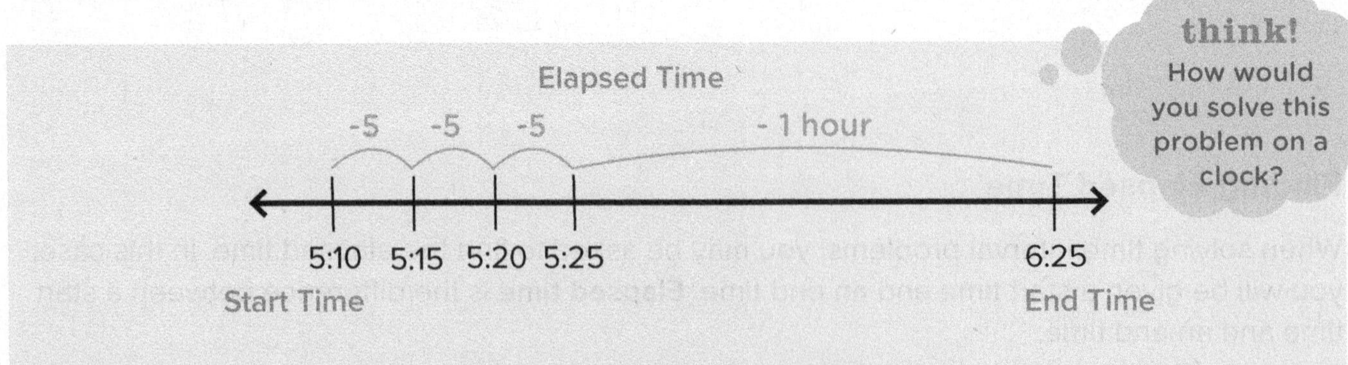

Julius got to practice at 5:10 p.m.

PRACTICE: Now you try

1. Molly woke up from a nap at 7:35 p.m. She had been asleep for 45 minutes. What time did she go to sleep?

   Start time ____________ Elapsed Time ____________ End Time ____________

2. Harry started eating dinner at 6:10 p.m. He finished at 6:58 p.m. How long did it take him to eat?

   Start time ____________ Elapsed Time ____________ End Time ____________

When Robert got to the library to study, he worked on a math assignment for 30 minutes, worked on a science project for 25 minutes, and studied for a spelling test for 15 minutes. He left the library at 4:30 p.m. What time did he get there? Show your work and explain your thinking on a piece of paper.

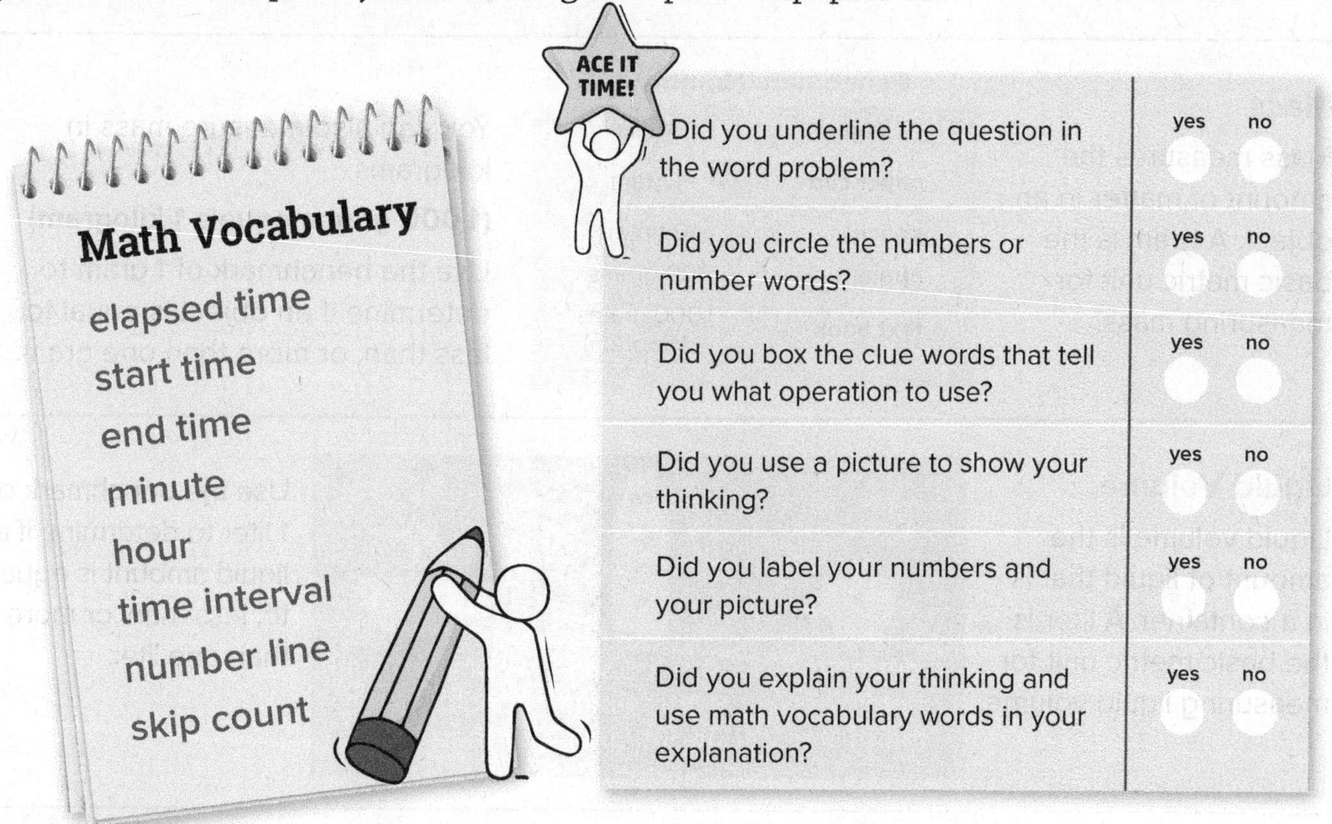

Math on the Move

To practice solving time interval problems, ask an adult or a friend to make up story problems about start times, end times, and elapsed times. Calculate the math to figure out the times.

Length, Mass, and Liquid Volume

FOLLOWING THE OBJECTIVE
You will practice measuring length to the nearest quarter- and half-inch. You will also measure and estimate liquid volumes and masses of objects using grams, kilograms, and liters.

LEARN IT: You can use tools to measure length, mass, and liquid volume of objects.

Length

Think of the parts of your ruler as a fraction or a number line.

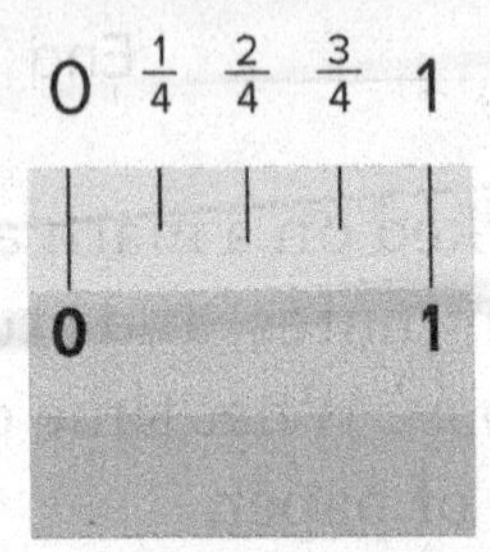

Divide the inch into two equal parts. This gives a $\frac{1}{2}$ marking on your ruler. Use this to measure an object to the nearest half-inch. If you divide the inch into four equal parts, there will be fourths markings on your ruler. You can use this to measure an object to the nearest quarter-inch.

Mass

Mass measures the amount of matter in an object. A gram is the basic metric unit for measuring mass.

Benchmark Numbers

Object	Mass
paper clip	1 gram
pencil	10 grams
apple	100 grams
text book	1,000 grams (1 kilogram)

You can also measure mass in kilograms.

(1,000 grams equals 1 kilogram)

Use the benchmark of 1 gram to determine if an object is equal to, less than, or more than one gram.

Liquid Volume

Liquid volume is the amount of liquid that is in a container. A liter is the basic metric unit for measuring liquid volume.

Use the benchmark of 1 liter to determine if a liquid amount is equal to, less than, or more than one liter.

PRACTICE: Now you try

1. Measure the object to the nearest half-inch.

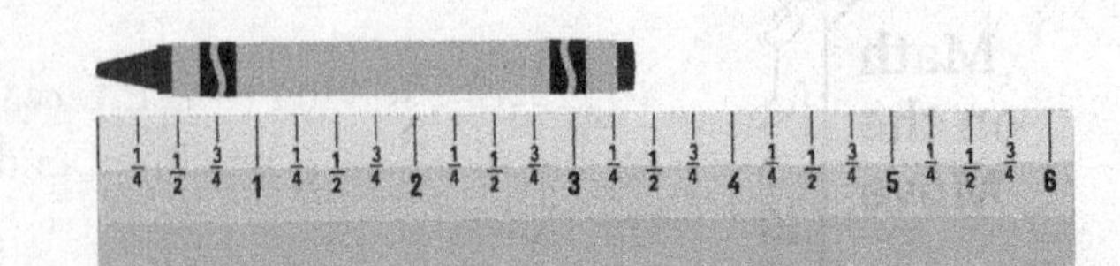

_______________ inches

2. Circle the unit you would use to measure the mass.

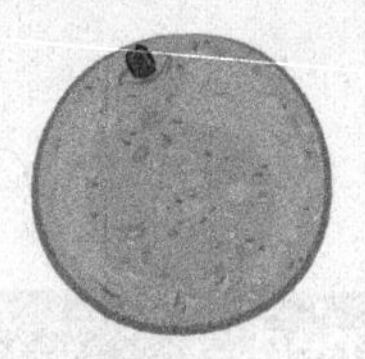

gram kilogram

3. Circle the best estimate of the liquid volume.

less than 1 liter

about 1 liter

more than 1 liter

Mrs. Fletcher baked a batch of 12 cookies. The total mass of all of the cookies was 84 grams. What was the mass of each cookie? Show your work and explain your thinking on a piece of paper.

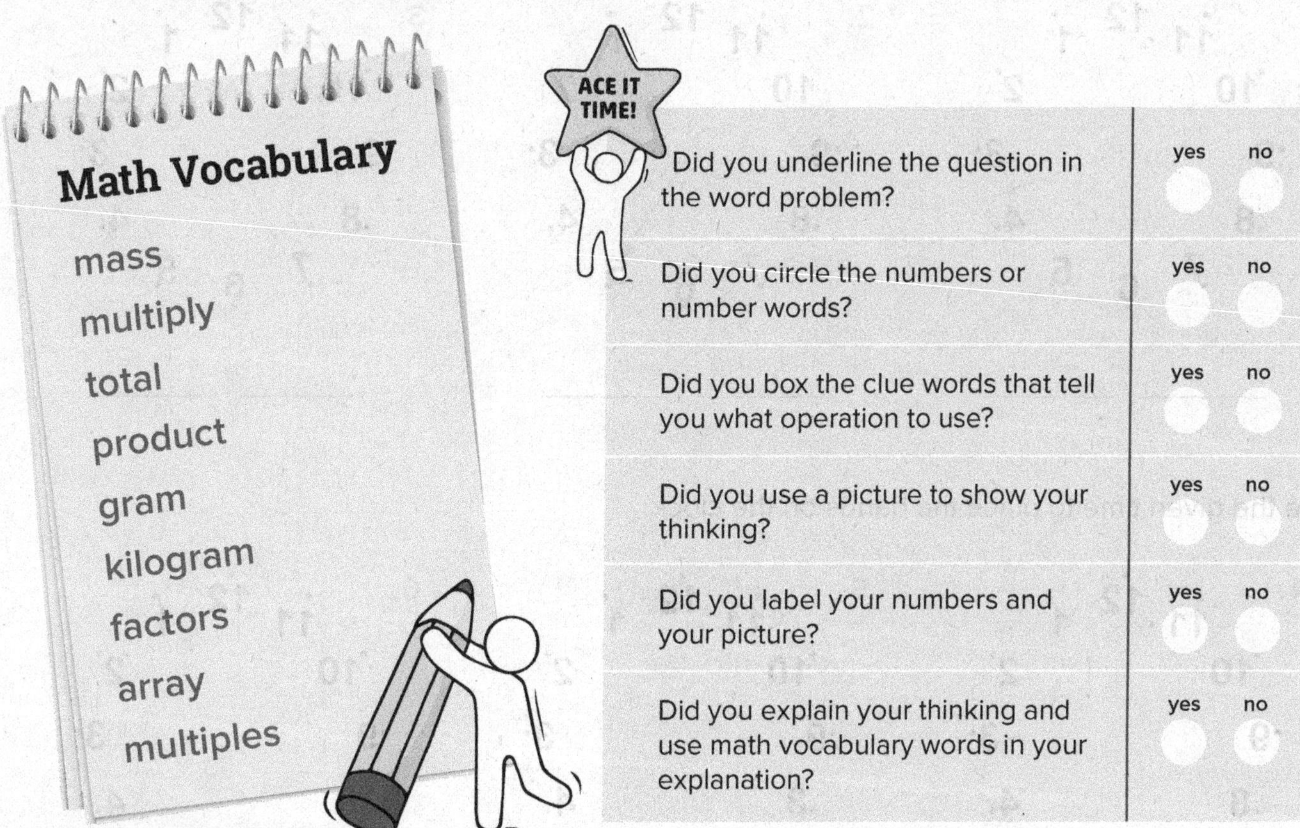

Math on the Move

Practice estimating the measurement of objects around the house. Measure length to the nearest quarter-inch and half-inch using a ruler. Estimate whether an object is less than, about equal to, or more than a gram, a kilogram, or a liter (using the table on page 112).

REVIEW

Stop and think about what you have learned.

Congratulations! You've finished the lessons for this unit. This means you have practiced telling and writing time. You have solved word problems about time intervals. You've learned how to measure and estimate liquid volumes and masses of objects. You've even learned to generate measurement data by using a ruler to measure length.

Now it's time to prove your measurement skills. Solve the problems below! Use all of the methods you have learned.

Activity Section

Write the time shown on the clock.

1. ______ : ______

2. ______ : ______

3. ______ : ______

Use the given time to place the hands on the clock.

4. 7:11

5. 2:23

6. 5:48

Solve the problems below. Draw a number line or clock to solve.

7. Dawn's piano rehearsal started at 12:45 p.m. She was done rehearsing at 2:55 p.m. How long was she in rehearsal?

   Start time ____________ Elapsed Time ____________ End Time ____________

8. Thea got home from the movies at 3:00 p.m. The movie was 2 hours and 30 minutes long, and it took her ten minutes to walk home from the movies. What time did the movie start?

   Start time ____________ Elapsed Time ____________ End Time ____________

9. Adam left his house at 2:15 p.m. to go to the store. It would normally take him 25 minutes to get to the store, but he made a stop at his grandmother's house for 15 minutes. What time was it when he finally got to the store?

   Start time ____________ Elapsed Time ____________ End Time ____________

10. Kelly went over to Lisa's house to hang out. They played games for 45 minutes, watched TV for 30 minutes, and worked on an art project for 20 minutes. Kelly left Lisa's house at 7:25 p.m. What time did Kelly get to Lisa's house?

    Start time ____________ Elapsed Time ____________ End Time ____________

11. Mrs. Oliver finished cleaning her house at 5:00 p.m. She started at 3:45 p.m. How long did it take Mrs. Oliver to clean her house?

    Start time ____________ Elapsed Time ____________ End Time ____________

12. The Mueller family got to the park at 12:35 p.m. They were there for 2 hours and 45 minutes. What time did the Muellers leave the park?

    Start time ____________ Elapsed Time ____________ End Time ____________

Measure objects below to solve.

13. Measure the pen to the nearest quarter-inch.

__________ inches

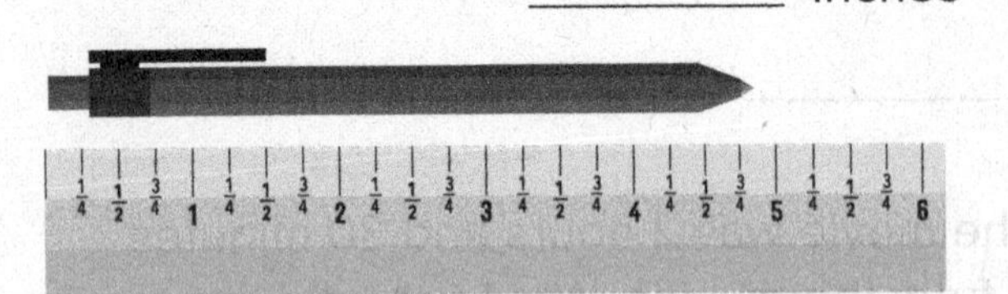

14. Measure the toy tool box to the nearest half-inch.

__________ inches

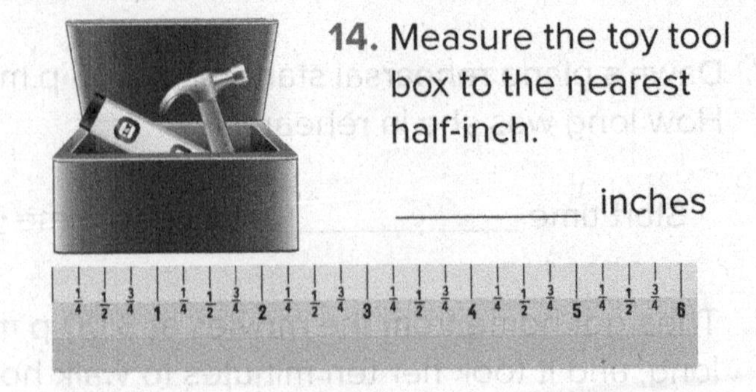

15. Measure the tube to the nearest quarter-inch.

__________ inches

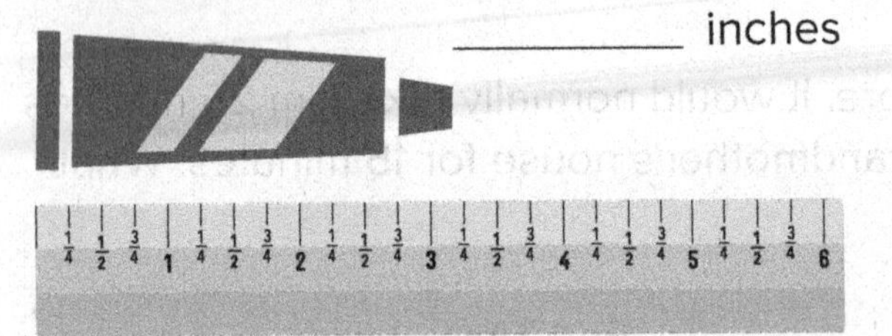

16. Measure the comb to the nearest half-inch.

__________ inches

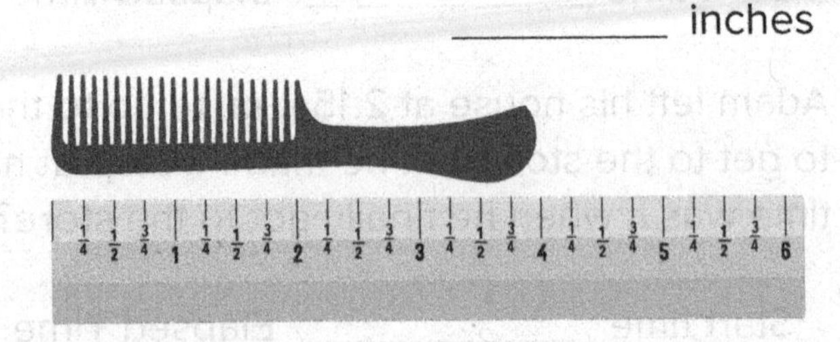

Circle the unit you would use to measure the mass.

17. gram / kilogram	**18.** gram / kilogram	**19.** gram / kilogram
20. gram / kilogram	**21.** gram / kilogram	**22.** gram / kilogram

Circle the best estimate of the liquid volume.

23.

less than 1 liter

about 1 liter

more than 1 liter

24.

less than 1 liter

about 1 liter

more than 1 liter

25.

less than 1 liter

about 1 liter

more than 1 liter

UNDERSTAND

Understand the meaning of what you have learned and apply your knowledge.

You understand how to generate measurement data by using a ruler to measure length to the nearest fourth or half of an inch.

Activity Section

Locate the following items around your home, including a ruler, and measure each to the **nearest half-inch**.

Measure the length of a spoon. ________ inches	Measure the length of a piece of mail. ________ inches
Measure the height of a water bottle. ________ inches	Measure the length of a book. ________ inches
Measure the length of a pencil. ________ inches	Measure the height of a coffee mug. ________ inches

Locate the following items around your home, including a ruler, and measure each to the **nearest half-inch**.

Measure the length of a cell phone. ________ inches	Measure the length of a child's shoe. ________ inches
Measure the length of a comb. ________ inches	Measure the height of a medicine bottle. ________ inches
Measure the length of a toothbrush. ________ inches	Measure the height of a remote control. ________ inches

Your family is moving to a new home! Now it is time to start packing. Your mom gives you a box that is 12 inches high. On the lines below, list some items (under 12 inches in height) in your home that could fit in the box so that the box can be covered by a lid.

________________________ ________________________

________________________ ________________________

DISCOVER

Discover how you can apply the information you have learned.

You can show that you understand how to tell and write time to the nearest minute. You can solve word problems involving addition and subtraction of time intervals in minutes.

Activity Section

Doug and his brother are taking a trip to another state on the bullet train. The train leaves at 1:00 a.m. They must be at the train terminal 45 minutes before the train leaves. The drive to the train terminal takes about 25 minutes. When they get to the terminal, they have to park their car in a parking garage and take a shuttle over to the train (this will take 20 minutes). What time should they leave their home? Is the time a.m. or p.m.? Show your work and explain your answer.

Perimeter and Area Concepts

What Is Perimeter?

FOLLOWING THE OBJECTIVE
You will practice solving problems involving the perimeter of polygons.

LEARN IT: ***Perimeter*** **is the distance around a shape.**

Finding Perimeter

You can estimate and measure perimeter using standard units like inches and centimeters.

A **polygon** is a closed plane figure with straight sides.

In order to find the perimeter of a polygon, you must find the sum of the length of its sides.

Let's find the perimeter of this rectangle:

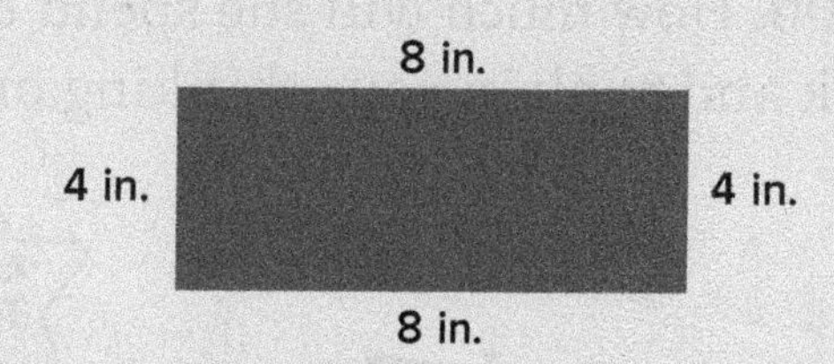

Add the lengths of each side together.

$8 + 4 + 8 + 4 = P$ ← symbol or variable for perimeter

$12 + 12 = 24$

You may also use grid paper to find the perimeter of a polygon by counting the number of units on each side.

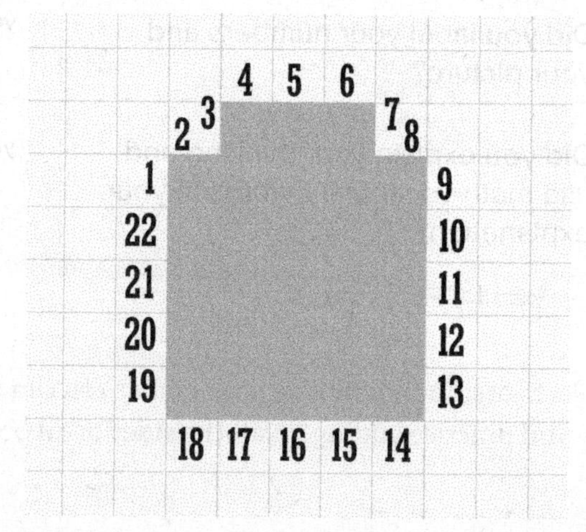

Start at a point and count the units around the shape until you have counted them all. Keep track by writing the numbers.

PRACTICE: Now you try

Find the perimeter of the shape. Each unit is 1 cm.

1.

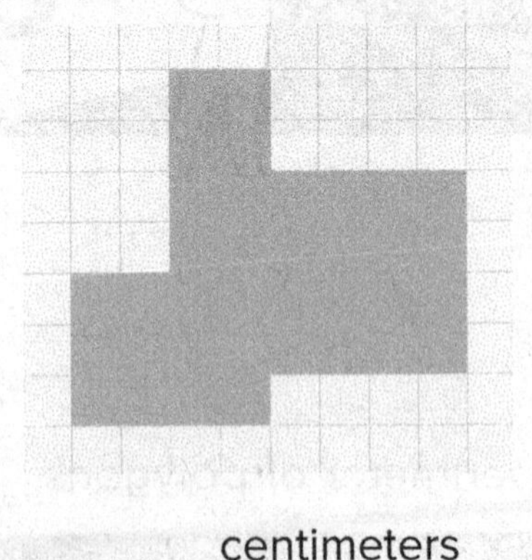

_______ centimeters

2.

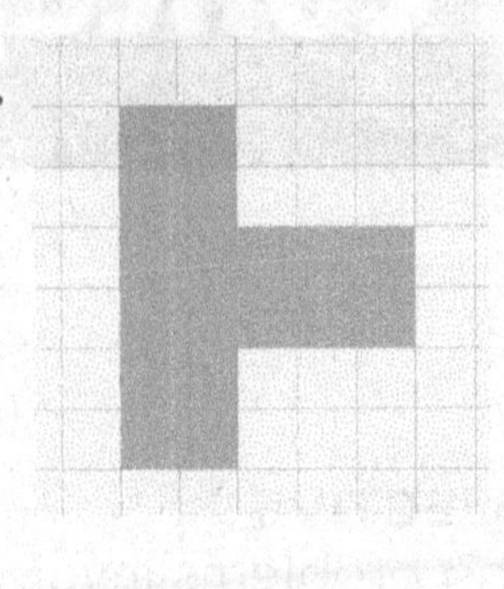

_______ centimeters

3.

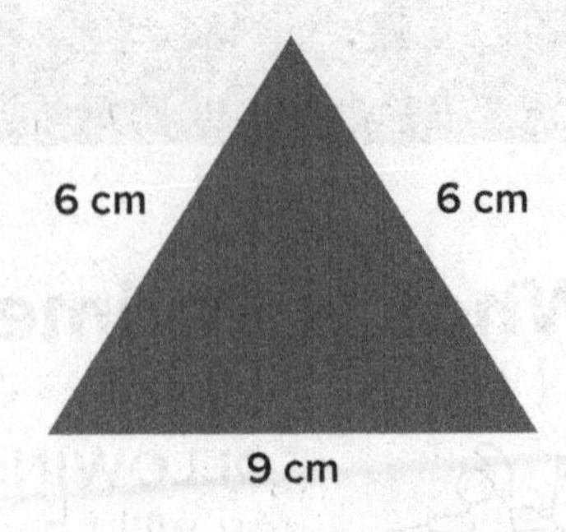

_______ centimeters

Carl's mom wants to put a fence around the garden. The family has a small vegetable garden. The rectangular garden is 15 feet long and 10 feet wide. Each foot of fencing costs $2.00. How much will she spend to put a fence up around the garden? Show your work and explain your thinking on a piece of paper.

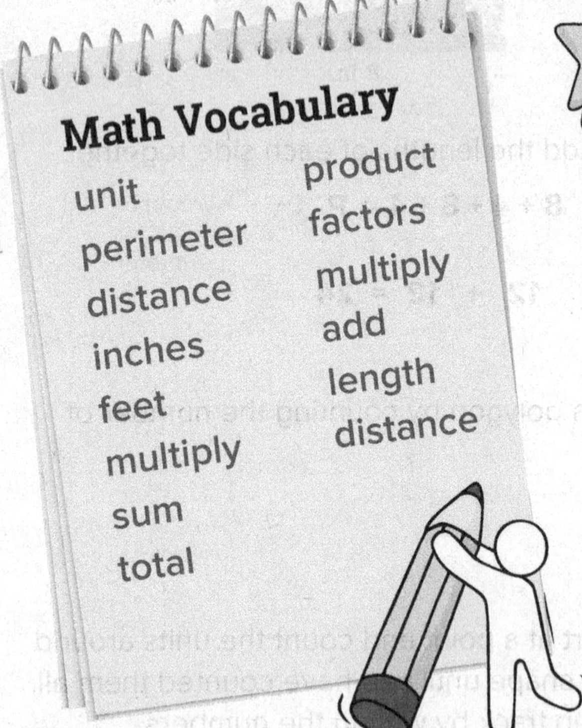

ACE IT TIME!

	yes	no
Did you underline the question in the word problem?		
Did you circle the numbers or number words?		
Did you box the clue words that tell you what operation to use?		
Did you use a picture to show your thinking?		
Did you label your numbers and your picture?		
Did you explain your thinking and use math vocabulary words in your explanation?		

Practice measuring the perimeter of objects inside and outside your home. You can use a ruler or other selected tools to measure and practice adding up all the sides of an object to find its perimeter.

Using Perimeter to Find Missing Sides

FOLLOWING THE OBJECTIVE
You will practice solving problems involving the perimeter of polygons.

LEARN IT: Find the missing length of a side of a polygon using what you know about perimeter.

Finding Perimeter

You can solve for the missing side of a polygon by using a given perimeter.

In order to find the missing side of a polygon, you must find the sum of the length of the sides given, and then subtract that sum from the total perimeter.

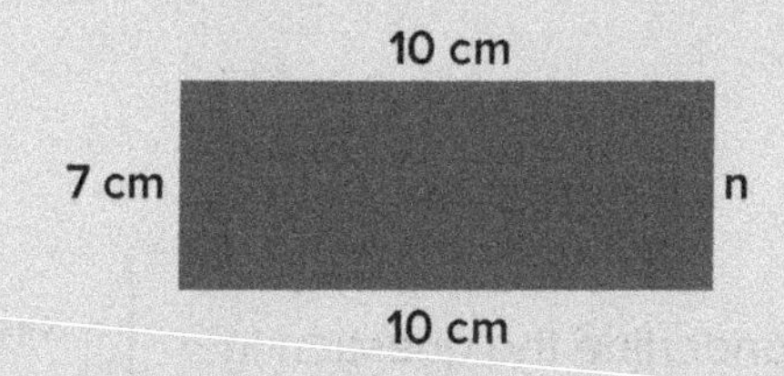

Let's find the missing side (n) of this rectangle. This rectangle has a perimeter of 34 cm.

1. Add the lengths of the given sides.
   10 + 7 + 10 = 27
2. Subtract the sum of the given sides from the perimeter. 34 – 27 = **7 cm**

You may also look at it this way:

10 + 7 + 10 + n = 34 ⟵ perimeter

27 + n = 34

34 – 27 = n Use the opposite operation to help solve for a variable or missing number.

7 cm = n Subtract 27 from 34.

PRACTICE: Now you try

Find the length of the missing side.

1. Perimeter = 36 centimeters

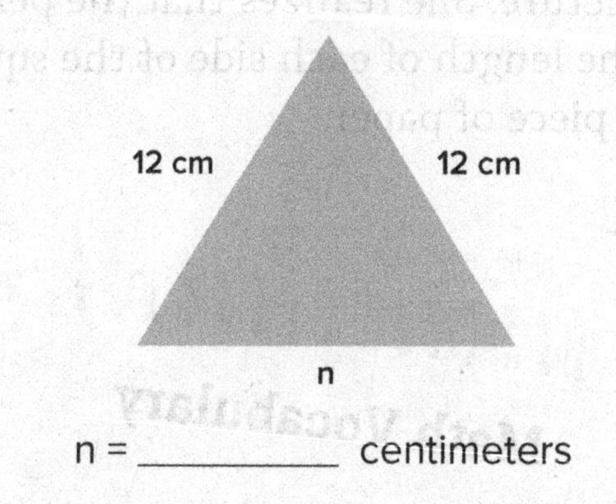

n = __________ centimeters

2. Perimeter = 48 inches

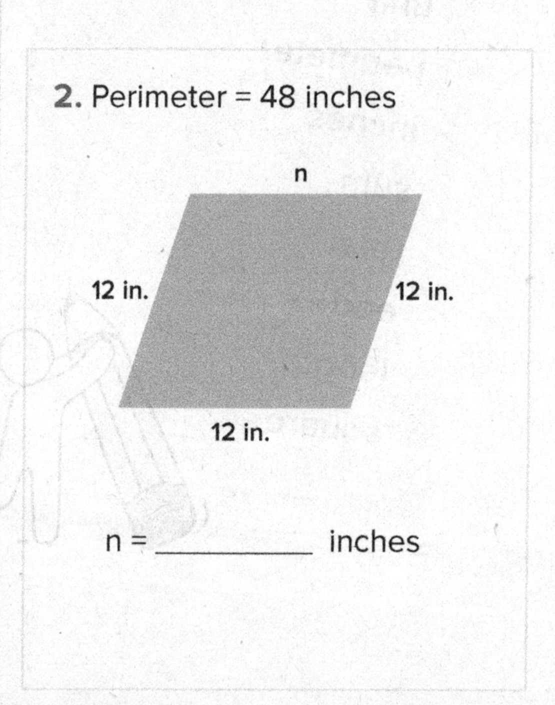

n = __________ inches

3. Perimeter = 116 inches

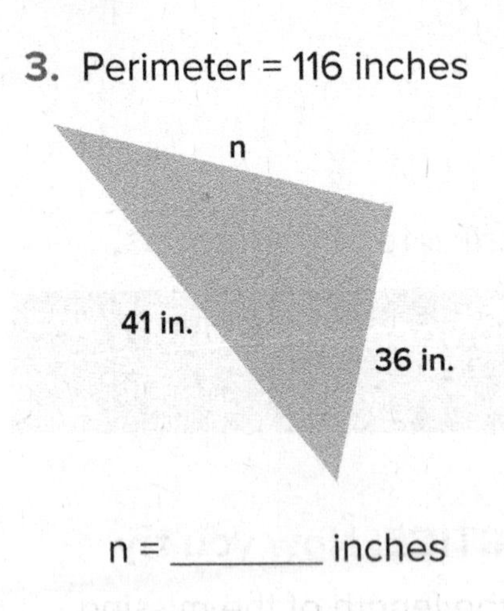

n = ________ inches

4. Perimeter = 70 centimeters

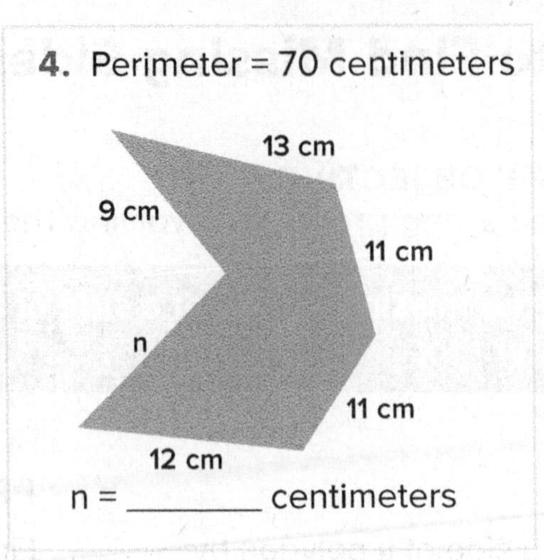

n = ________ centimeters

5. Perimeter = 59 centimeters

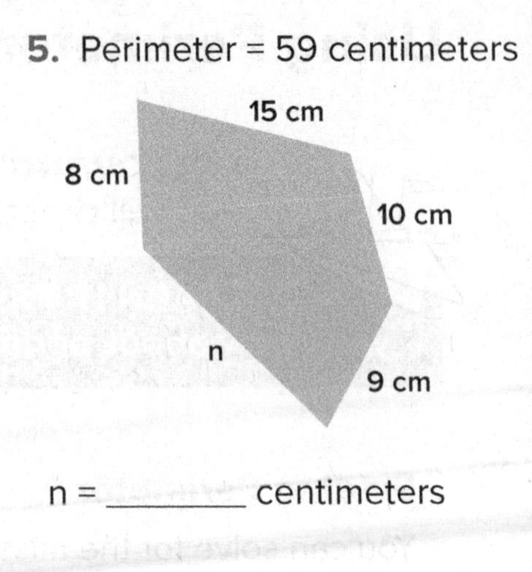

n = ________ centimeters

Olivia is making a picture for her mom. She wants to put a ribbon border around the picture. She realizes that the perimeter of the square picture is 36 inches. What is the length of each side of the square? Show your work and explain your thinking on a piece of paper.

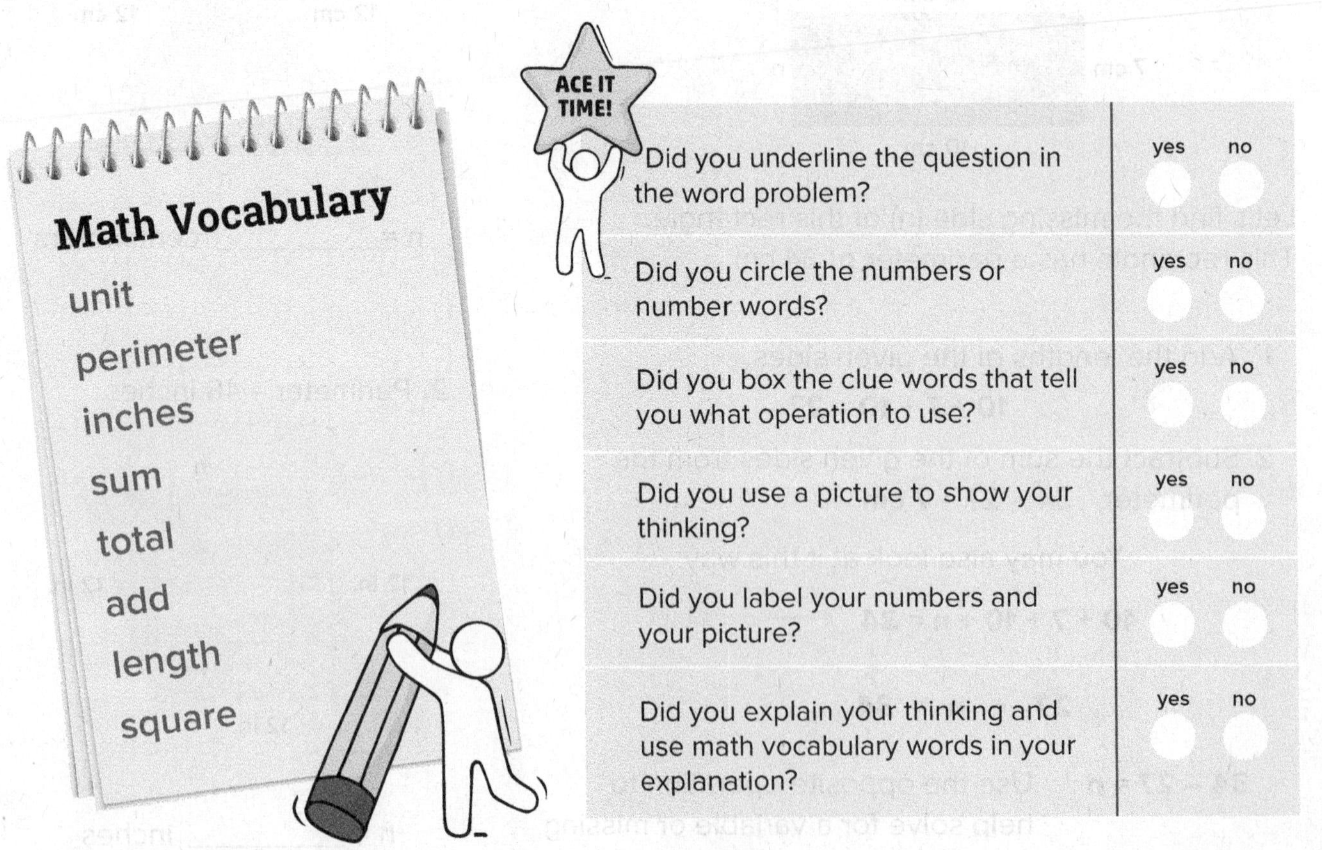

What Is Area?

FOLLOWING THE OBJECTIVE
You will understand that area can be measured by the number of square units it takes to cover (without gaps or overlaps) a plane figure.

LEARN IT: *Area* is the measure of the number of unit squares needed to cover a flat surface. The difference between perimeter and area is that perimeter is the distance around an object to enclose it. Area is the space within the object.

Understanding Area

A **unit square** can be used to find the area of a plane (flat) figure. A unit square has one square unit of area.

AREA 1 unit 1 unit

A unit square has four sides that each measure 1 unit long.

A unit square can be represented by any unit used to measure length (e.g., centimeters, inches, feet, yards, meters, miles). So, a unit square can equal 1 square centimeter, 1 square foot, or even 1 square mile (pick the appropriate unit depending on what is being measured).

You can find the area of a plane figure by calculating how many unit squares it will take to cover that figure without any gaps or overlaps. Let's look at this rectangle.

How many unit squares will it take to cover this rectangle without any gaps or overlaps?

It will take **10** unit squares of this size to cover this rectangle.

PRACTICE: Now you try

Count the unit squares to find the area of each shape.

1.

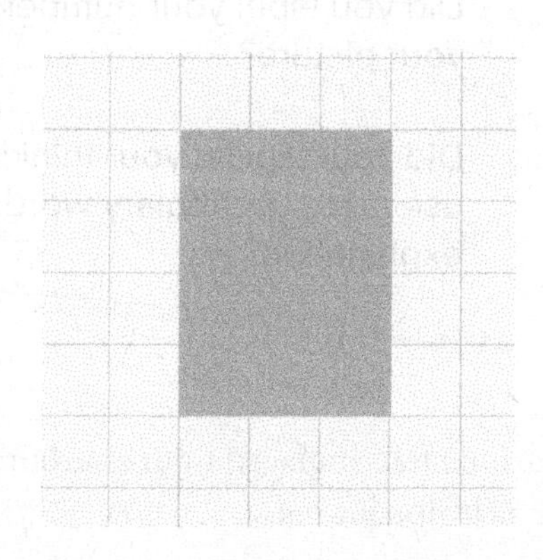

Area = ________ square units

2.

Area = ________ square units

3. Draw a shape with an area of 12 square units.

Count to find the area of each shape.

4. Area = ____ square units

5. Area = _____ square units

Robert and Chris are working on an art project together. They use a piece of cardboard that is shaped like a rectangle. The side lengths are 5 units and 3 units. Robert says that the area of their cardboard is 16 square units, and Chris says that the area of their cardboard is 15 square units. Who is correct? Show your work and explain your thinking on a piece of paper.

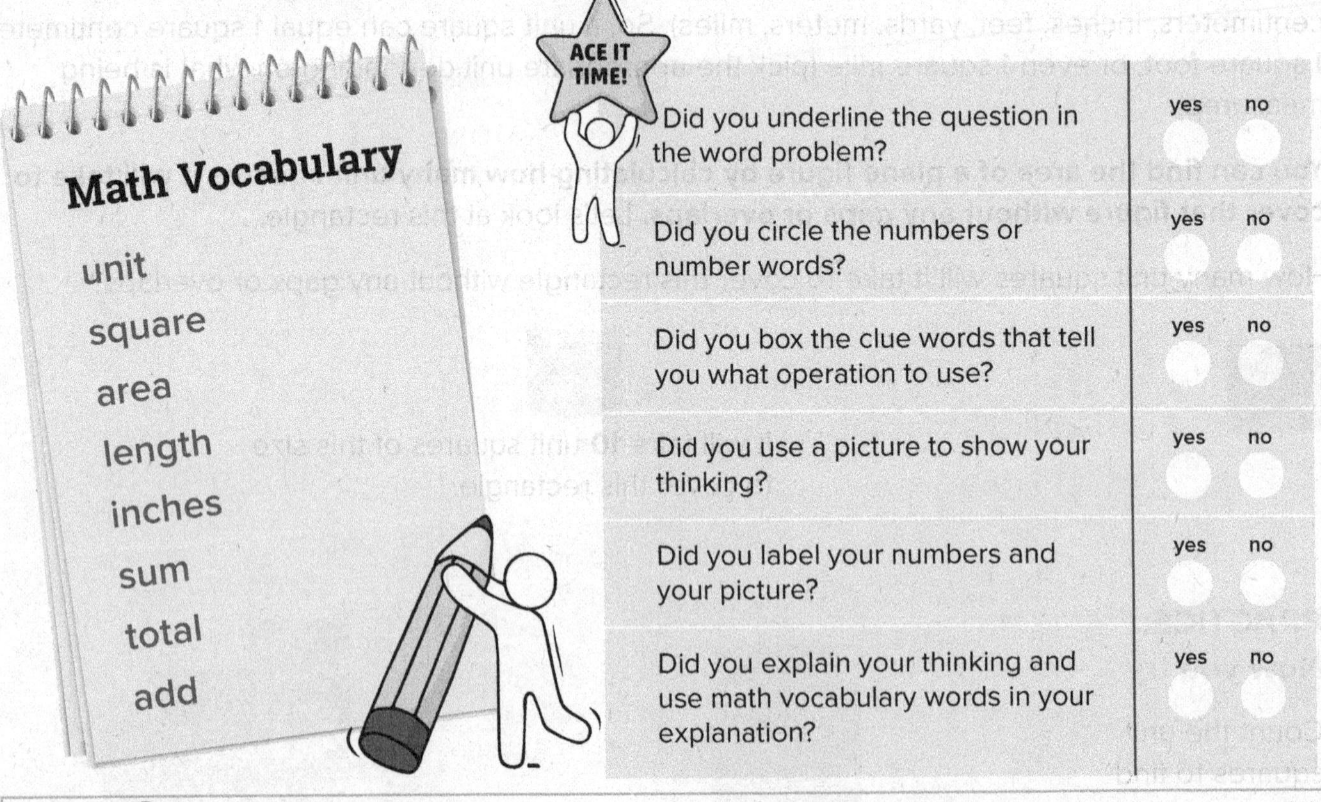

Practice measuring the area of various flat-surfaced figures around your home by using different sizes of square crackers as unit squares.

More Work with Area

FOLLOWING THE OBJECTIVE
You will practice tiling and multiplying sides and lengths to solve for area.

LEARN IT: You can find the area of a rectangle by tiling or by multiplying.

Remember: A unit square can be represented by any unit used to measure length (e.g., centimeters, inches, feet, yards, meters, miles). So a unit square can equal 1 square centimeter, 1 square foot, or even 1 square mile (pick the appropriate unit depending on what is being measured).

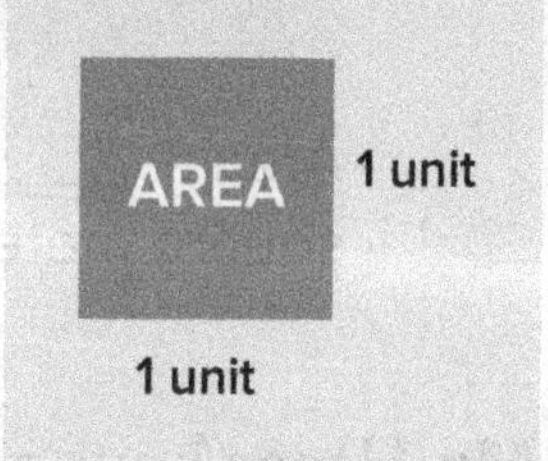

Let's use tiling to find the area of Zoey's bathroom rug.

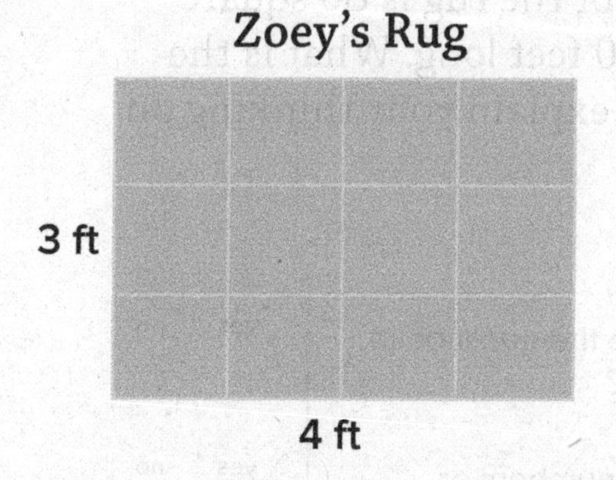

Zoey's bathroom rug measures 4 **feet** by 3 **feet**. We will tile by covering the area with unit squares, or drawing lines to make unit squares that represent **1-foot squares**. We know that our unit squares will represent 1-foot squares because the length of the sides of the rug is measured in feet. Now count the unit squares that cover the top of her rug.

It takes 12 (1-foot) unit squares to cover Zoey's rug, so the area of Zoey's rug is 12 square feet.

Let's use what we know about multiplication to find the area of Zoey's bathroom rug.

Look at the picture of Zoey's tiled bathroom rug. This looks like an array. You learned about arrays in Unit 4. Arrays help you multiply. You can use multiplication to find area.

In this case, there are three rows of tiles and there are four tiles in each row (three groups of four). Write this as an equation: **3 × 4 = 12**

PRACTICE: Now you try

Draw lines or multiply to find the area.

1.
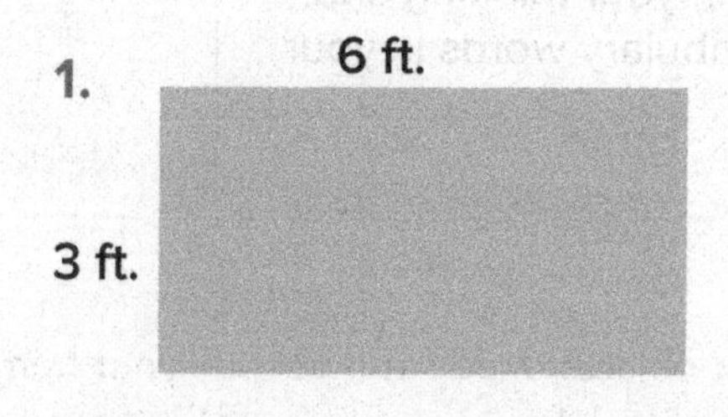

Area = ________ square feet

2.
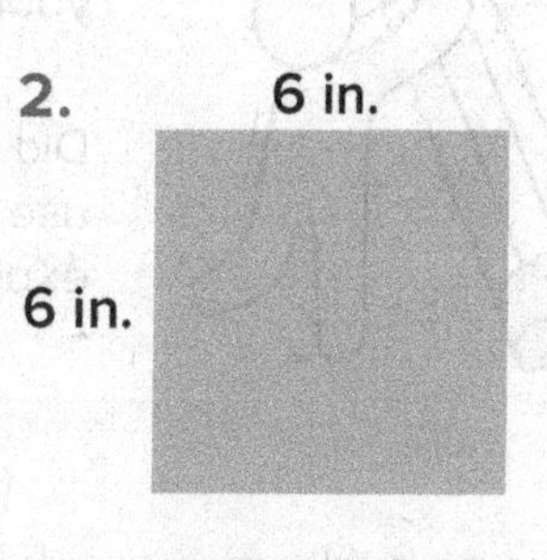

Area = ________ square inches

3.
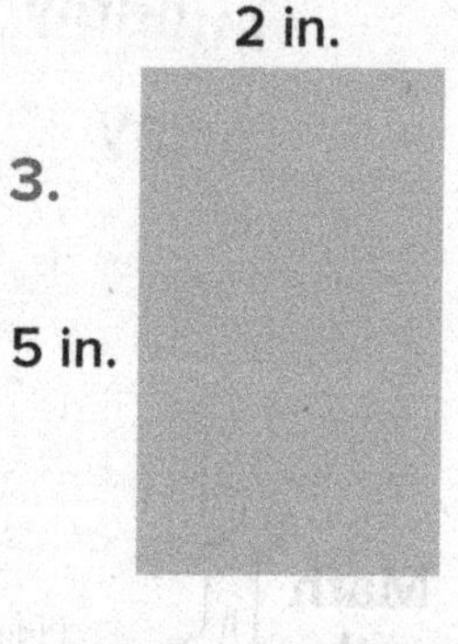

Area = ________ square inches

Use what you know about multiplication and arrays to solve for area.

4.

Area = ________ square meters

5. Andrea is sewing a quilt for her bed. The quilt is shaped like a rectangle. The lengths of the sides of the quilt measure 7 feet by 5 feet. What is the area of the quilt?

Area = ________ square feet

Mr. Hines buys an area rug for his living room. The area of the rug is 80 square feet. He knows that the length of one side of his rug is 10 feet long. What is the length of the other side of the rug? Show your work and explain your thinking on a piece of paper.

Math Vocabulary

- unit square
- area
- length
- feet
- total
- multiply
- array

	yes	no
Did you underline the question in the word problem?		
Did you circle the numbers or number words?		
Did you box the clue words that tell you what operation to use?		
Did you use a picture to show your thinking?		
Did you label your numbers and your picture?		
Did you explain your thinking and use math vocabulary words in your explanation?		

Math on the Move

Use various square units to measure the area of flat objects inside and outside your home. Remember you need to cover the entire area by repeating the square unit without any gaps or overlapping.

Use Smaller Rectangles to Find Area

FOLLOWING THE OBJECTIVE

You will practice solving for area by using the Distributive Property of Multiplication and breaking apart figures into smaller rectangles.

LEARN IT: Let's look at some different ways to find the area of rectangles.

Example: Find the area of the rectangle below using the Distributive Property.

The sides of this rectangle measure 11 inches and 3 inches. This rectangle is an array with 3 rows of 11 units, or 3 groups of 11. You can use the Distributive Property to break that array apart.

think!
We can break apart 11 into 8 and 3. Then we can distribute the 3 factor to both the 8 and the 3!

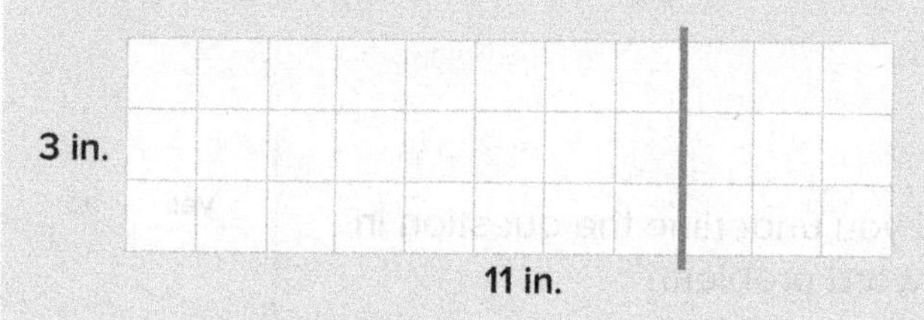

$$3 \times 11 = 3 \times (8 + 3)$$
$$= (3 \times 8) + (3 \times 3)$$
$$= 24 + 9$$
$$= 33$$

The area of this rectangle is 33 square inches.

Now look at the figure below. How can we find the area of this figure? We can break it apart into two smaller rectangles! Find the area of the smaller rectangles and then add them together to find the area of the total shape.

Step 1: Draw a line to break apart the shape into two rectangles. Label each rectangle.

Step 2: Find the area of each rectangle.

Area of Rectangle A = 2 × 6, or 12 square cm

Area of Rectangle B = 3 × 2, or 6 square cm

Step 3: Add the area of both rectangles to find the total area of the figure.

12 + 6 = 18

So, the area of the figure is 18 square centimeters.

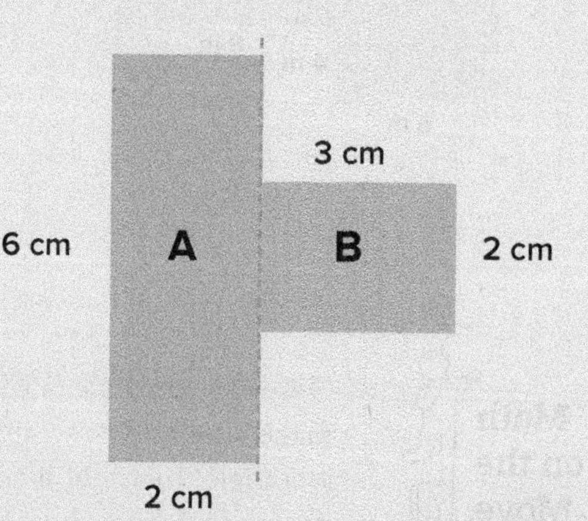

PRACTICE: Now you try

1. Use the Distributive Property to find the area.

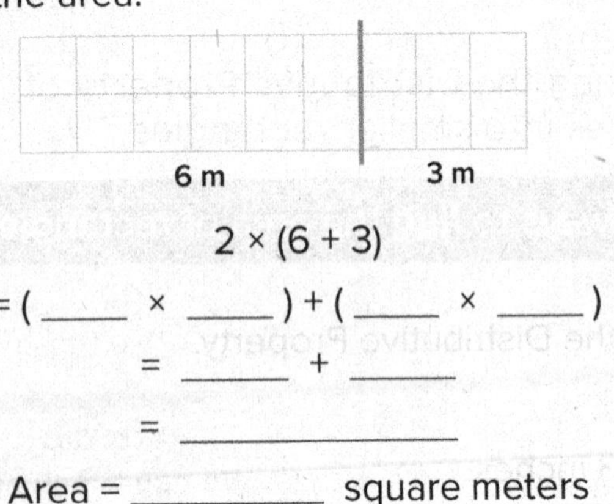

2 × (6 + 3)

= (_____ × _____) + (_____ × _____)

= _____ + _____

= _____________

Area = __________ square meters

2. Use the Break Apart Method to find area.

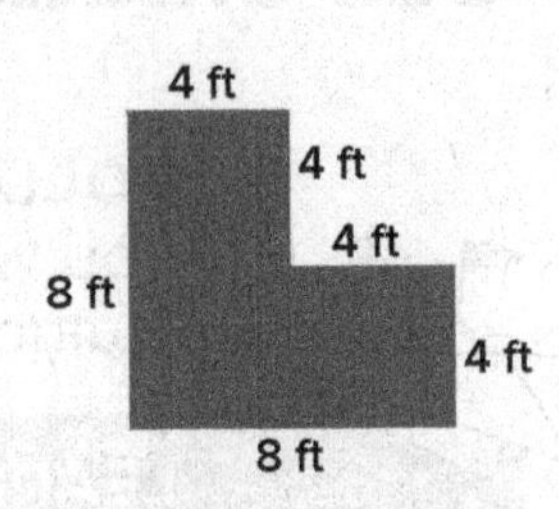

Rectangle A: _____ × _____ = _____

Rectangle B: _____ × _____ = _____

_____ + _____ = _____ square cm

Area of figure: _____ square cm

Sheila broke apart a figure into rectangles to find its area. She found that the area of the figure is 120 square meters. Do you agree with her? Why or why not? Show your work and explain your thinking on a piece of paper.

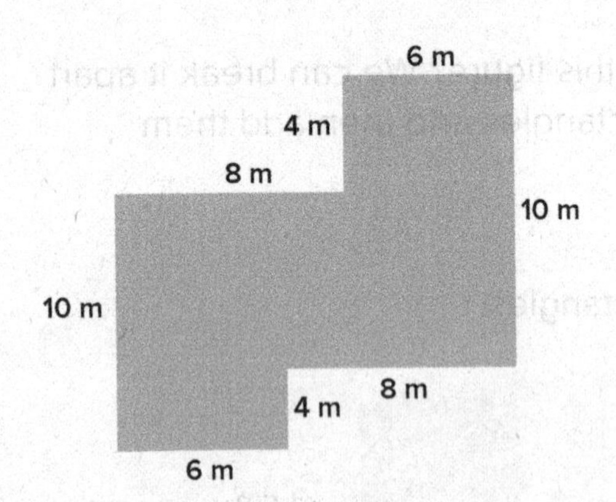

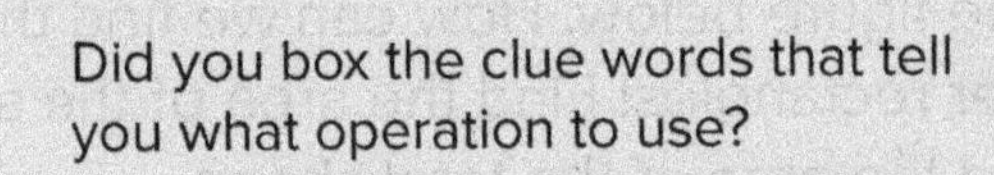

ACE IT TIME!

	yes	no
Did you underline the question in the word problem?		
Did you circle the numbers or number words?		
Did you box the clue words that tell you what operation to use?		
Did you use a picture to show your thinking?		
Did you label your numbers and your picture?		
Did you explain your thinking and use math vocabulary words in your explanation?		

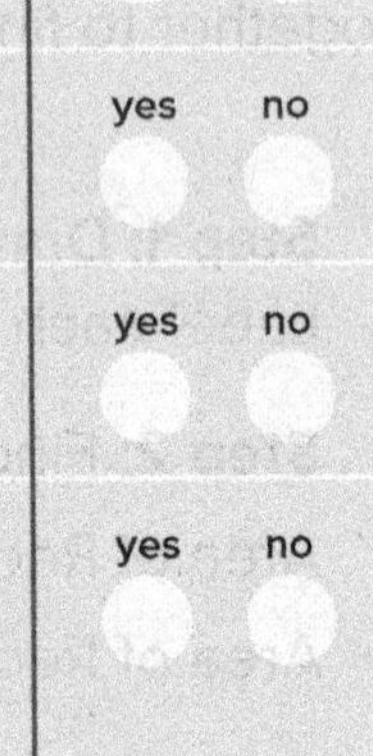

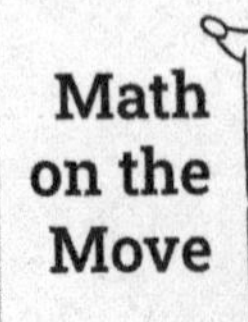

Math on the Move

Cut out a rectangle or square from a piece of graph paper. If you don't have graph paper, make a square or rectangle using a ruler to measure the sides. Find the area of the rectangle. Then, cut it apart to make two rectangles or squares. Find the area of the two new shapes. Show how the two areas added together will equal the sum of the original shape!

Relating Perimeter and Area

FOLLOWING THE OBJECTIVE

You will solve problems involving rectangles with the same perimeter and different areas, or rectangles with the same area and different perimeters.

LEARN IT: Sometimes two rectangles can have the same area with different perimeters. They can also have the same perimeter, but different areas. How can that be? Remember: Area is the number of units needed to cover a flat surface. It is measured in squared units. Perimeter is the distance around a shape.

Example: Compare the perimeter and the area of the rectangles below.

Step 1: Find the perimeter of each rectangle. ***Hint:*** Add up the sides to find the perimeter!

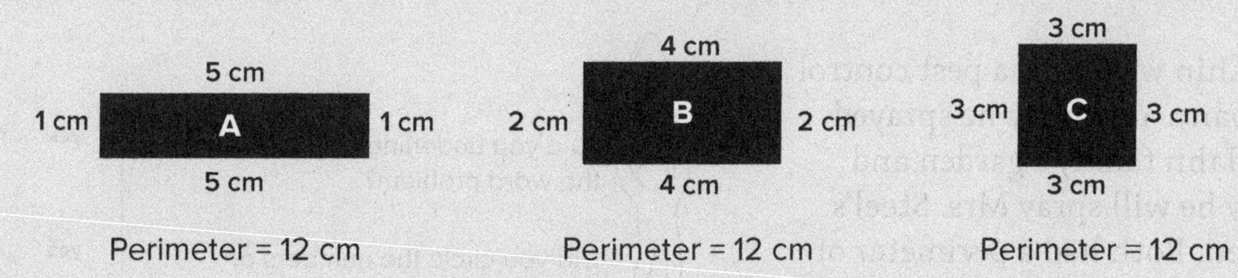

Perimeter = 12 cm | Perimeter = 12 cm | Perimeter = 12 cm

Step 2: Find the area of each rectangle.

Rectangle A	**Rectangle B**	**Rectangle C**
1 × 5 = 5	2 × 4 = 8	3 × 3 = 9
Area = 5 square centimeters	Area = 8 square centimeters	Area = 9 square centimeters

Step 3: Compare.

All three rectangles have a perimeter of 12, but all three have a different area!

Now let's look at the rectangles below. These rectangles have the same area but different perimeters!

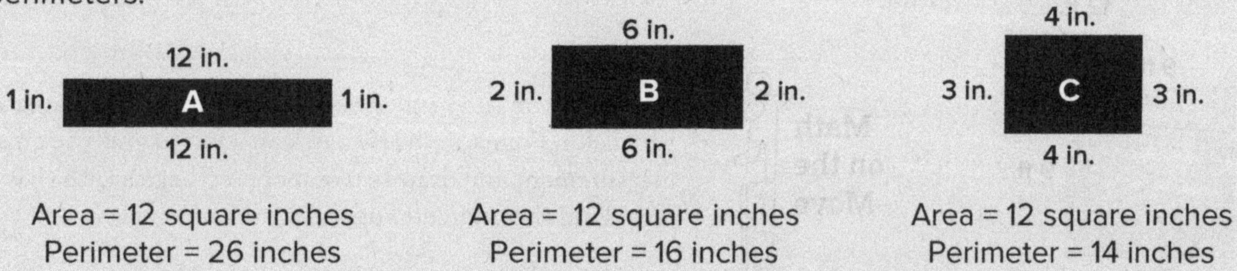

Rectangle A	Rectangle B	Rectangle C
Area = 12 square inches	Area = 12 square inches	Area = 12 square inches
Perimeter = 26 inches	Perimeter = 16 inches	Perimeter = 14 inches

PRACTICE: Now you try

Find the area and perimeter of each rectangle. Draw a different rectangle with the same area and a different perimeter in the box to the right.

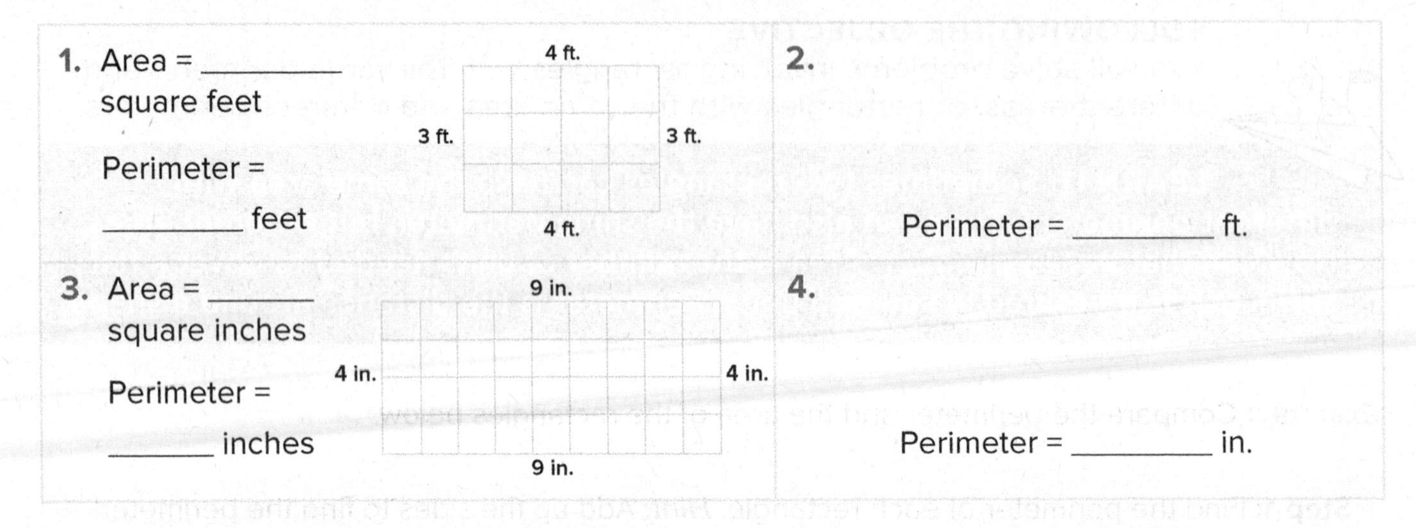

1. Area = ________ square feet
   Perimeter = ________ feet

2. Perimeter = ________ ft.

3. Area = ______ square inches
   Perimeter = ______ inches

4. Perimeter = ________ in.

Mr. Chin works for a pest control company. Yesterday he sprayed the Hahn family's garden and today he will spray Mrs. Steel's garden. Both had a perimeter of 36 feet. Look at Mr. Chin's paperwork. On which day did Mr. Chin spray the most bug spray in order to cover the area of the garden? Show your work and explain your thinking on a piece of paper.

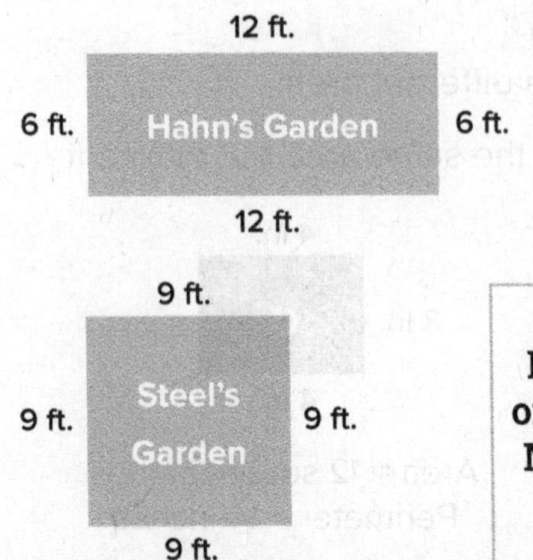

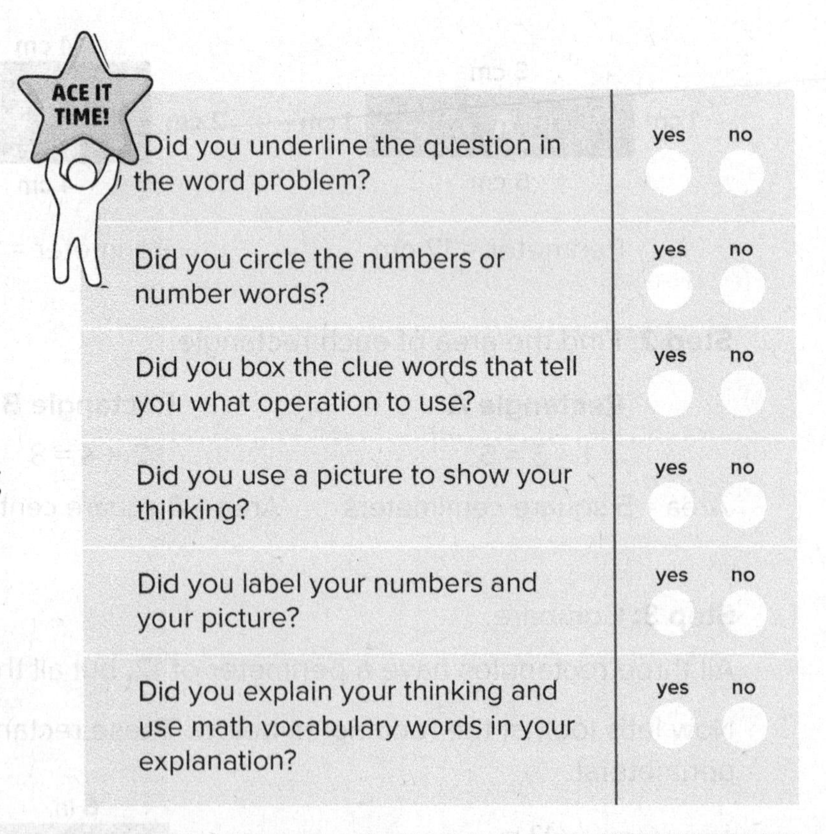

	yes	no
Did you underline the question in the word problem?		
Did you circle the numbers or number words?		
Did you box the clue words that tell you what operation to use?		
Did you use a picture to show your thinking?		
Did you label your numbers and your picture?		
Did you explain your thinking and use math vocabulary words in your explanation?		

Math on the Move

Practice estimating and measuring objects for their area and perimeter. Compare the two. Ask an adult to give you an area measurement and draw two or more rectangles with that area. Find the perimeter of each of the rectangles.

REVIEW

Stop and think about what you have learned.

Congratulations! You've finished the lessons for this unit. This means you've learned how to measure to solve for area. You can solve word problems involving perimeter. You can even make connections between perimeter and area.

Now it's time to prove your skills with perimeter and area. Solve the problems below! Use all of the methods you have learned.

Activity Section

Find the perimeter of the figure below.

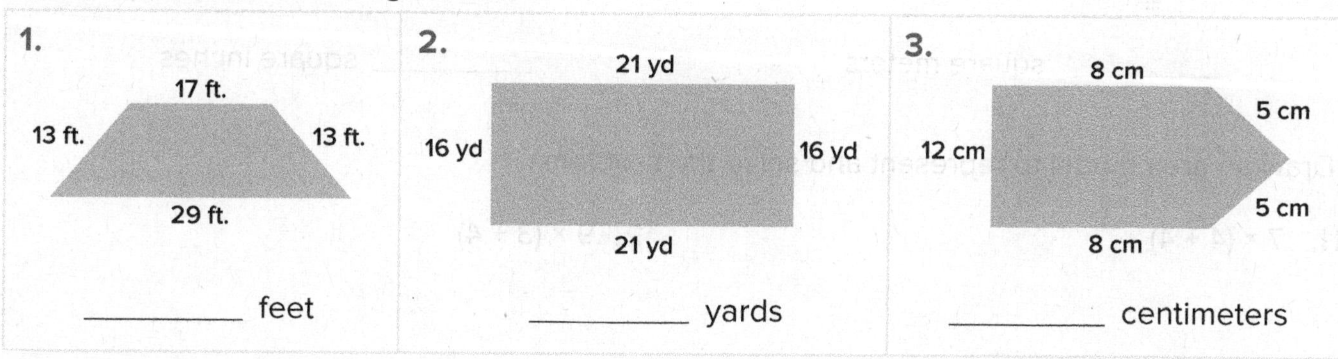

1. __________ feet

2. __________ yards

3. __________ centimeters

Use the given perimeter to solve for the missing side.

4. Perimeter = 24 ft.

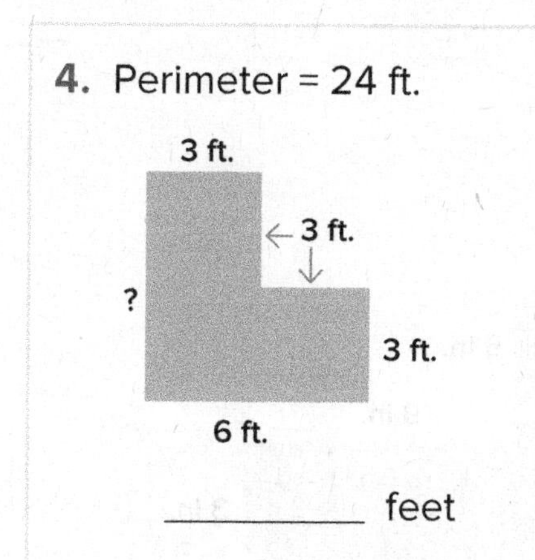

__________ feet

5. Perimeter = 67 cm

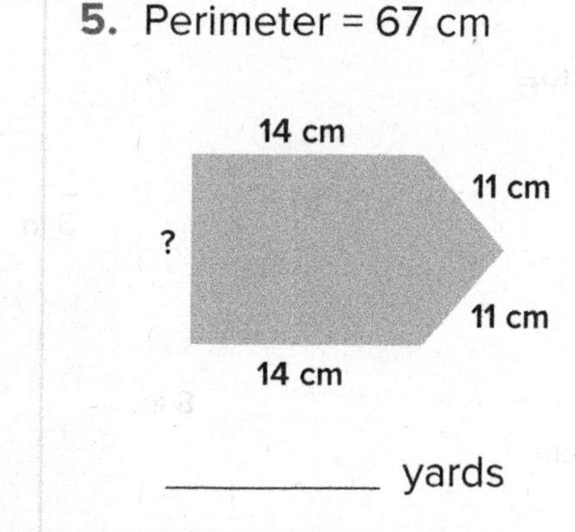

__________ yards

6. Perimeter = 80 in.

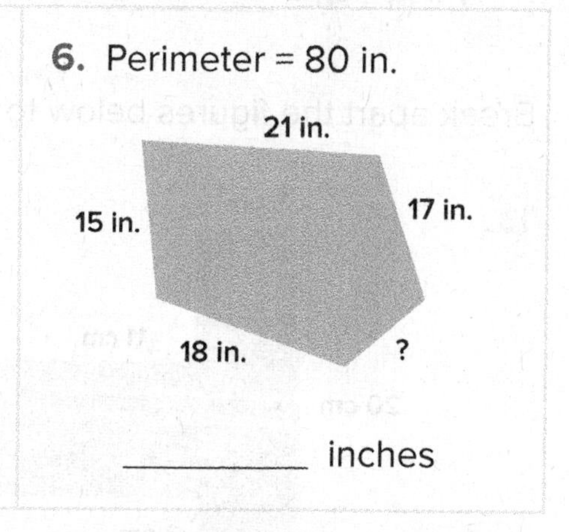

__________ inches

Tile each figure. Draw lines to make unit squares to cover each figure. Count to find the area.

7.

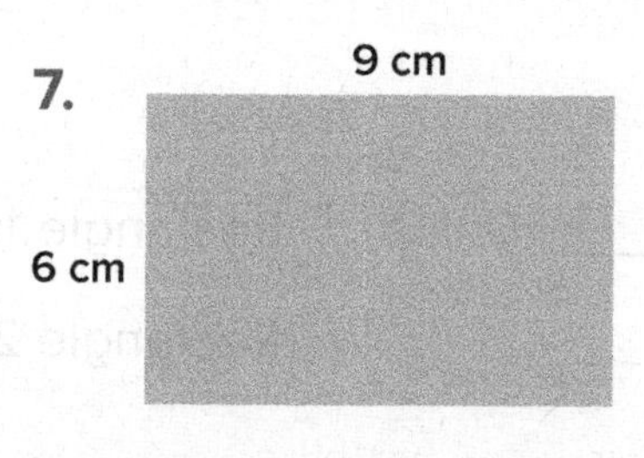

Area = __________ square centimeters

8.

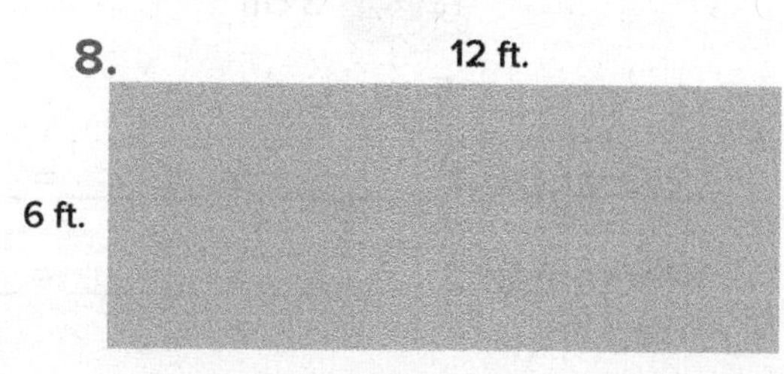

Area = __________ square feet

Use the Distributive Property to solve for area.

9.

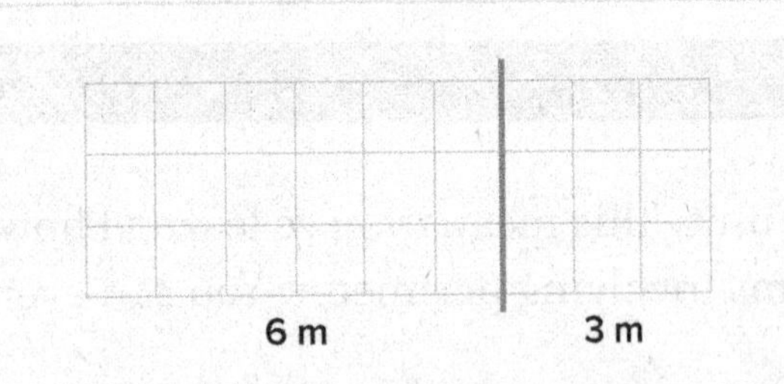

$3 \times (6 + 3)$

= (_____ × _____) + (_____ × _____)

= _____ + _____

= __________

_______ square meters

10.

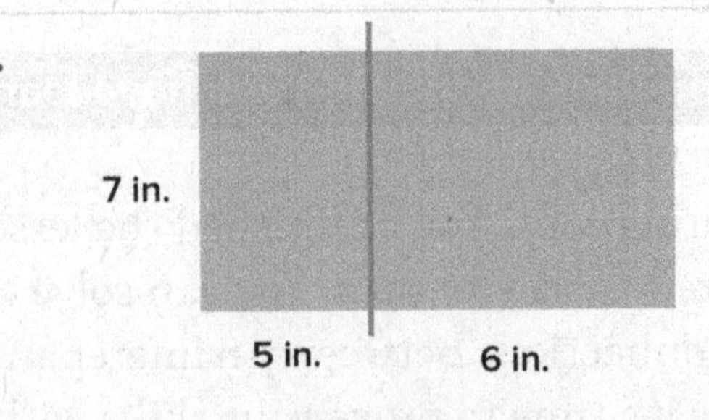

$7 \times (5 + 6)$

= (_____ × _____) + (_____ × _____)

= _____ + _____

= __________

_______ square inches

Draw an area model to represent and solve the problems.

11. $7 \times (4 + 4)$

$7 \times (4 + 4) =$ _______

12. $9 \times (3 + 4)$

$9 \times (3 + 4) =$ _______

Break apart the figures below to solve.

13.

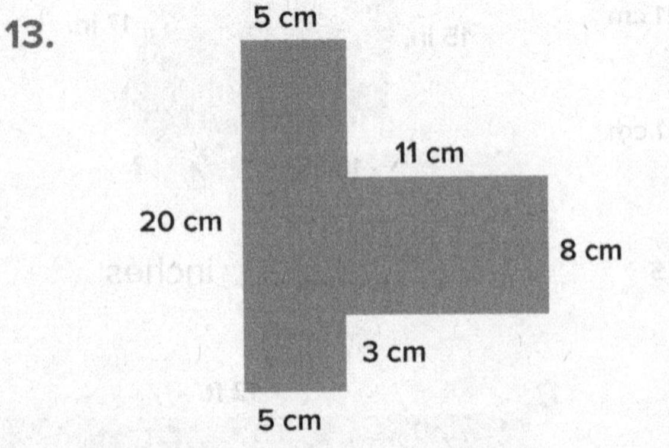

Rectangle 1: _____ × _____ = _____

Rectangle 2: _____ × _____ = _____

_____ + _____ = _____ square cm

14.

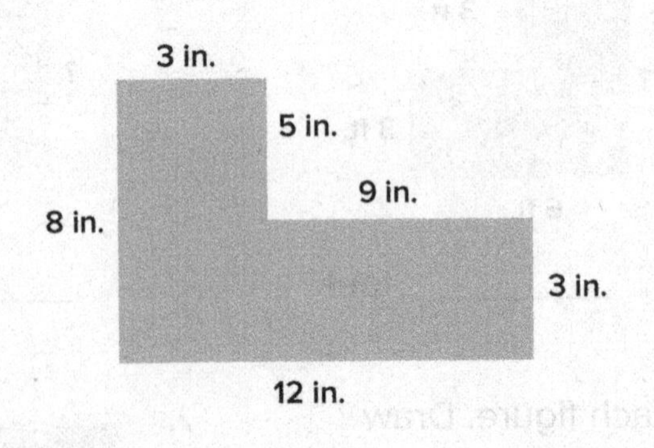

Rectangle 1: _____ × _____ = _____

Rectangle 2: _____ × _____ = _____

_____ + _____ = _____ square in.

15. Find the area of the given rectangle. Draw three different rectangles with the same perimeter and different areas in the three empty boxes. Label them.

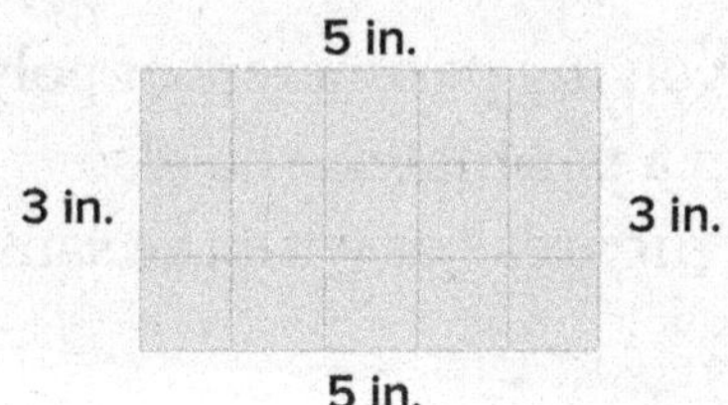

Area = ________ square in.

Rectangle 1: Perimeter = 16 in.

Area = ________ square in.

Rectangle 2: Perimeter = 16 in.

Area = ________ square in.

Rectangle 3: Perimeter = 16 in.

Area = ________ square in.

16. Find the area and perimeter of the given rectangle. Draw two different rectangles with the same area and different perimeters in the two empty boxes. Label them.

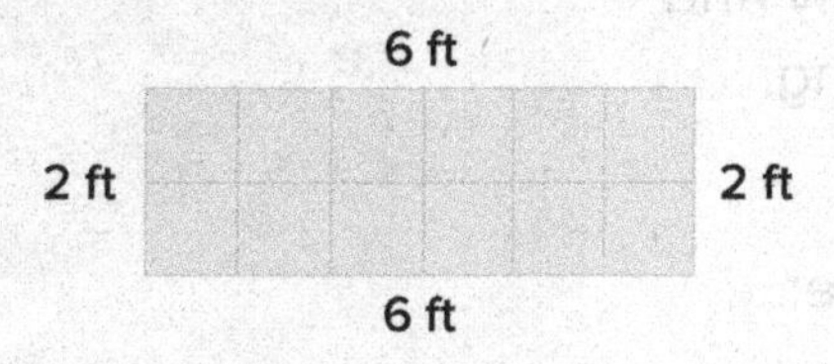

Perimeter = ________ feet

Area = ________ square feet

Rectangle 1: Area = 12 square ft.

Perimeter = ________ feet

Rectangle 2: Area = 12 square ft.

Perimeter = ________ square feet

UNDERSTAND

Understand the meaning of what you have learned and apply your knowledge.

You understand how to solve real-world math problems involving perimeters of polygons, including finding an unknown side of an object when given a perimeter. You also understand rectangles with the same perimeter and different areas or with the same area and different perimeters.

You understand how to multiply side lengths to find the area of rectangles in the context of real-life and mathematical problems.

You recognize that area is additive. You understand that the area of various shapes can be found by breaking apart the shape into rectangles, finding the area of these rectangles, and adding them to find the area of the whole figure.

Activity Section

Mr. Coleman adopted a community garden from the city. It is in bad condition and he wants to fix it up. The garden has fencing on three sides and is missing a piece of fence on one side. His first order of business is to replace the missing piece of fence. The city gave him a model of the property. He can tell from the model that the perimeter of the rectangular garden is 46 feet long. He looks closely and can make out that one side of the garden is 15 feet long.

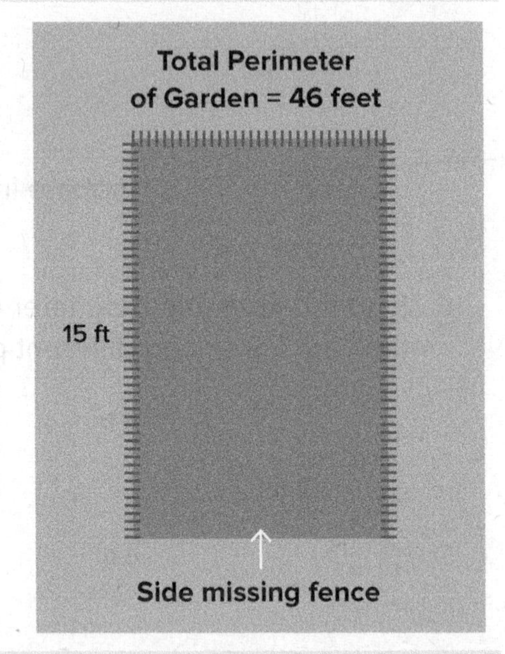

How many feet of fencing will he have to buy for the missing side? Show your work and explain your answer.

DISCOVER

Discover how you can apply the information you have learned.

You understand how to solve real-world math problems involving perimeters of polygons, including finding an unknown side of an object when given a perimeter. You also understand rectangles with the same perimeter and different areas or with the same area and different perimeters.

Activity Section

Your mom has asked you to help make a plan to arrange the tables at an upcoming dinner party. Seven guests will be joining your family of five. She has several square tables that can seat one person on each side (each table is one square unit). She has asked you to arrange them so in an array so that no tables are touching. All guests are to be seated. There should be no empty seats.

Draw a picture of how you would set up the tables and the chairs for the party.

How many tables did you use? __________

What is the area of your table arrangement? __________

Can you think of another way you could set up the tables for the party? Draw it.

How many tables did you use this time? __________

What is the area of your table arrangement? __________

Can you think of another way you could set up the tables for the party? Draw it.

How many tables did you use this time? __________

What is the area of your table arrangement? __________

Geometry Concepts

Shapes and Attributes

FOLLOWING THE OBJECTIVE

You will understand that shapes fit into different categories depending on their characteristics.

LEARN IT: Classify a polygon using its sides and angles.

Words to Know

Two-Dimensional Shape: A shape that has only two dimensions. It has length and width, but no thickness (i.e., circles, triangles, squares).

Polygon: A two-dimensional shape made up of only straight sides and angles. There must be at least three sides.

Two **sides** of a polygon meet or intersect at a vertex (point); this forms an angle.

Angles

Right Angle

An angle that makes a square corner. The angle measures exactly 90 degrees.

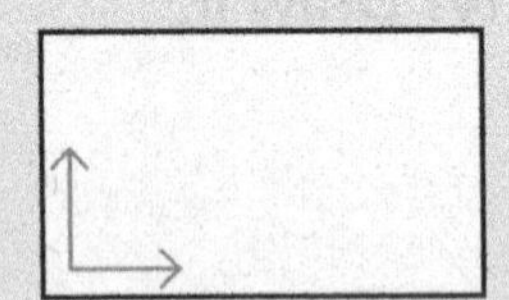

This is a right angle.

Acute Angle

An angle that is smaller than a right angle. The angle measures less than 90 degrees.

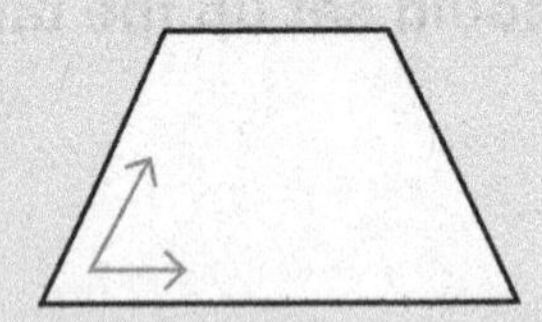

This is an acute angle.

Obtuse Angle

An angle that is greater than a right angle. The angle measures more than 90 degrees.

This is an obtuse angle.

Lines

Intersecting Lines

Lines that cross or meet to form angles.

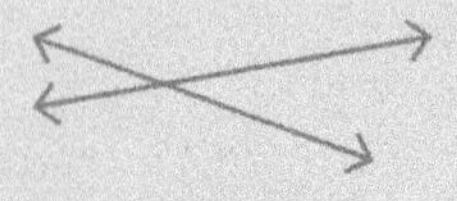

Perpendicular Lines

Intersecting lines that cross or meet to form right angles.

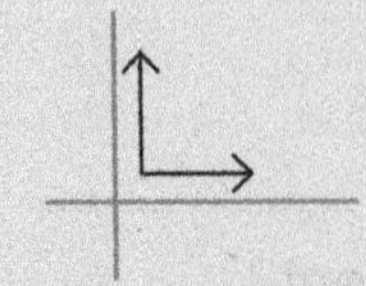

Parallel Lines

Lines that are the same distance apart. They never cross.

Classifying Polygons

Triangle: 3 sides, 3 angles

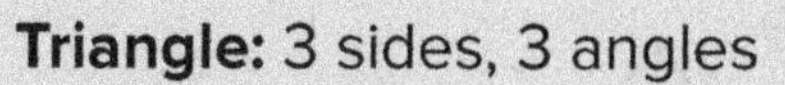

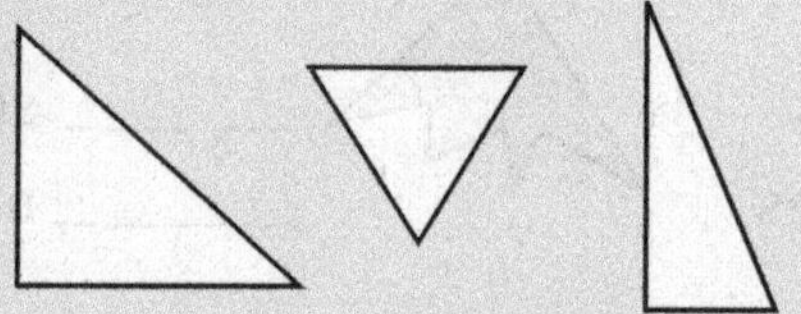

Quadrilateral: 4 sides, 4 angles

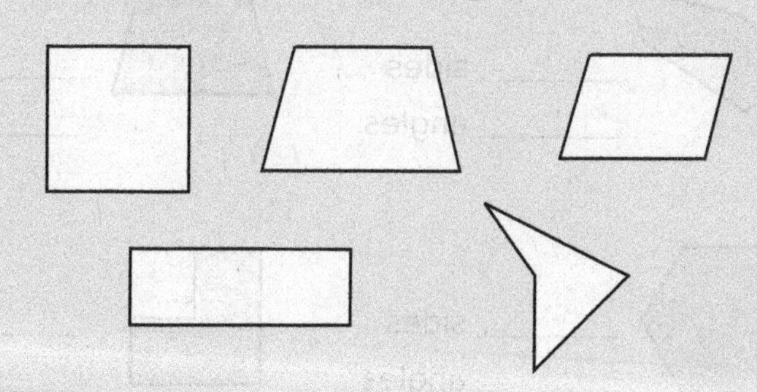

Pentagon: 5 sides, 5 angles

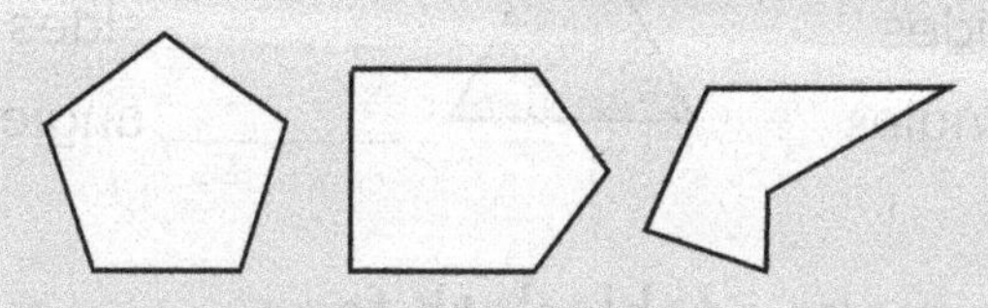

Octagon: 8 sides, 8 angles

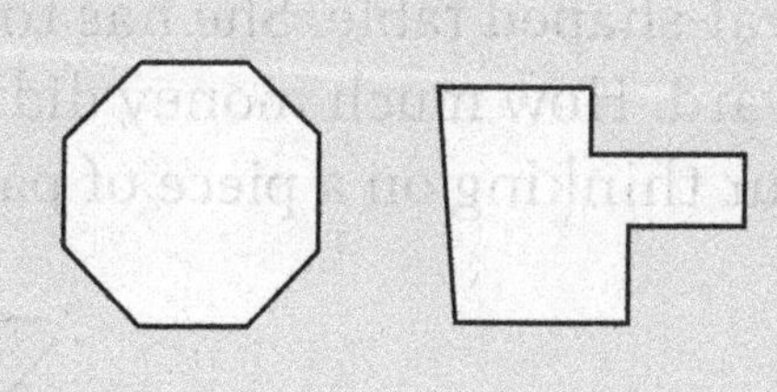

Hexagon: 6 sides, 6 angles

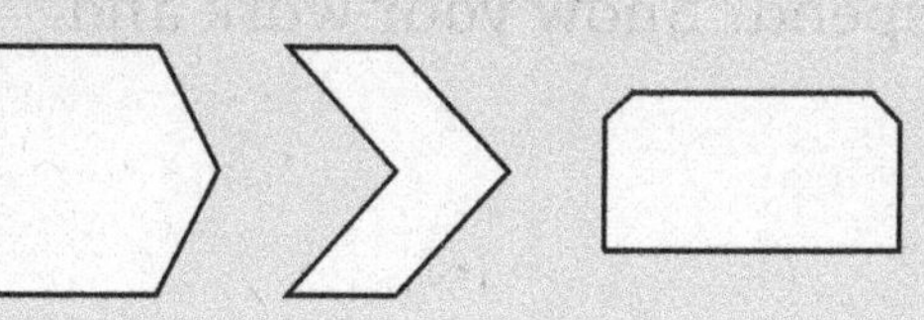

PRACTICE: Now you try

Write whether the angle is acute, obtuse, or right.

1. ____________________

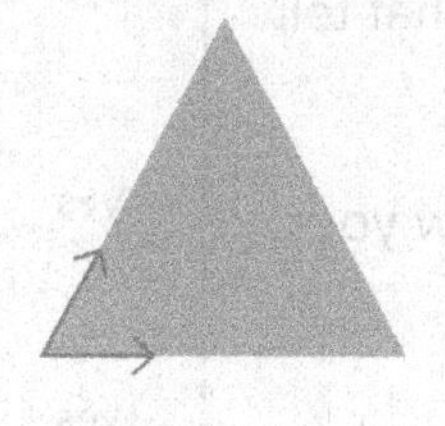

2. ____________________

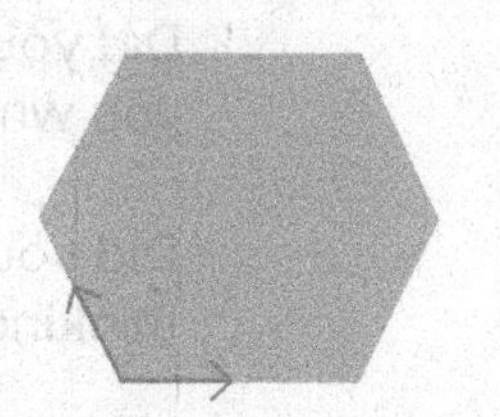

3. ____________________

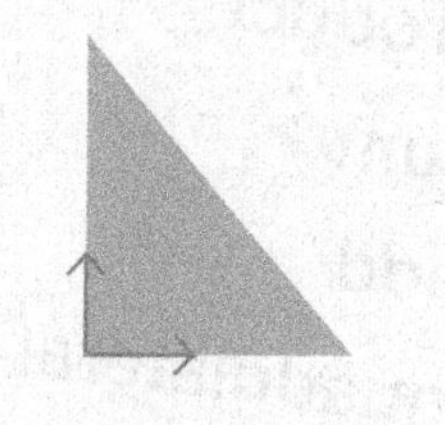

Do the sides of the polygon form parallel lines, intersecting lines, or perpendicular lines?

4. ____________________

5. ____________________

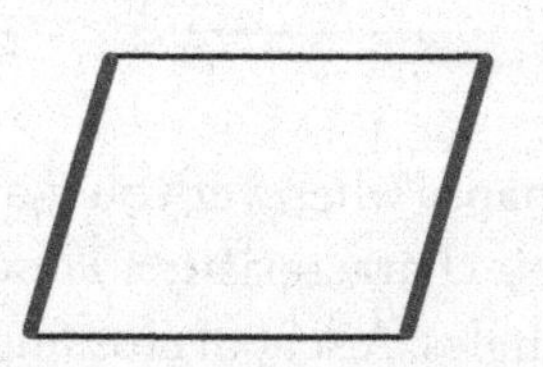

6. ____________________

List how many sides and how many angles are in each polygon.

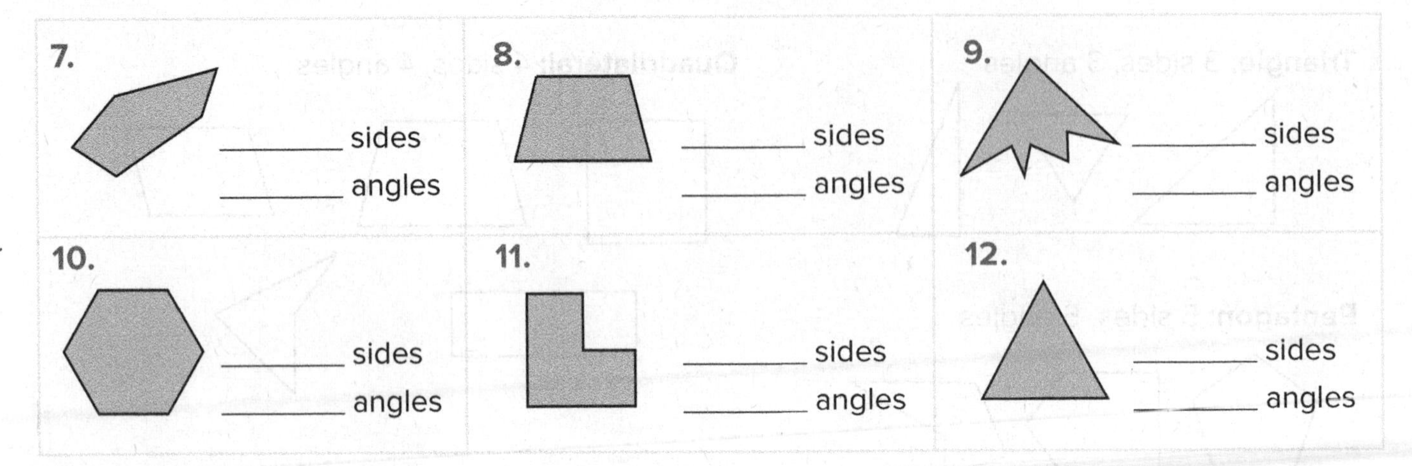

Mrs. Griffin goes to the fabric store to buy fabric to make a tablecloth for a quadrilateral-shaped table. She has to buy four yards for each side. The fabric costs $2.00 per yard. How much money did Mrs. Griffin spend? Show your work and explain your thinking on a piece of paper.

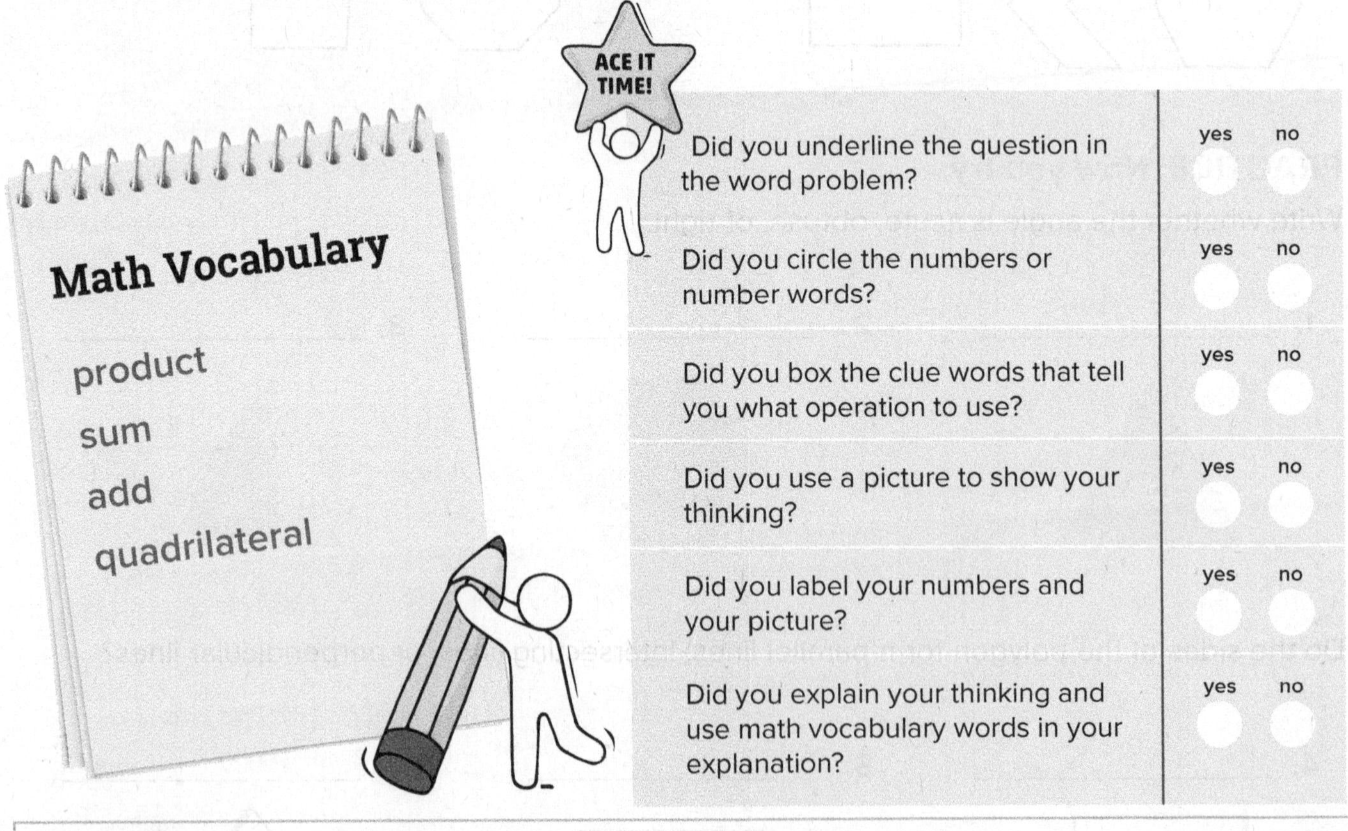

Math on the Move

Locate two-dimensional shapes wherever you may be. Practice naming the shapes and have discussions about their characteristics. For example, a stop sign is an octagon and it has eight sides and eight angles. A school crossing sign has five sides and five angles.

Quadrilaterals

FOLLOWING THE OBJECTIVE

You will understand that quadrilaterals fit into different categories depending on their characteristics. You will also practice drawing quadrilaterals.

LEARN IT: Practice identifying and drawing different types of quadrilaterals.

Quadrilaterals

Remember: A quadrilateral has four sides and four angles. Quadrilaterals can be classified by the length of their sides, the relationship of their sides, and the size of their angles.

Quadrilaterals (shapes with 4 sides and 4 angles)

Trapezoid
Exactly 1 pair of opposite parallel sides

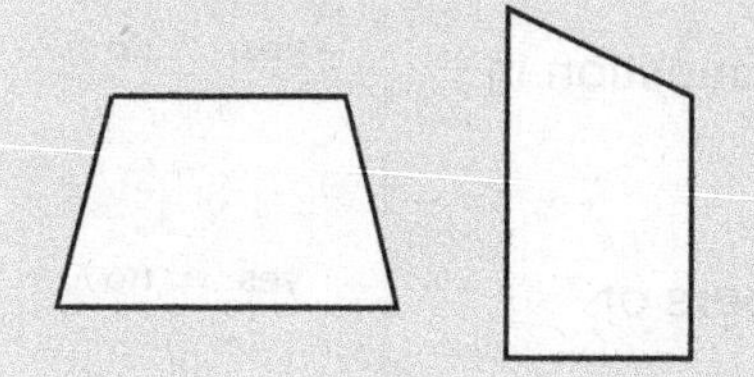

Parallelogram
2 pairs of opposite parallel sides
Lengths of opposite sides are equal

Other Quadrilaterals
Have 4 sides and 4 angles but do not fit the definition of trapezoid or parallelogram

A rhombus and a rectangle are both parallelograms.

Rhombus
2 pairs of opposite parallel sides
Lengths of opposite sides are equal
All sides are equal

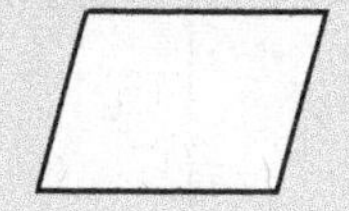
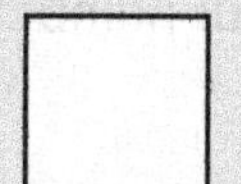

Rectangle
2 pairs of opposite parallel sides
Lengths of opposite sides are equal
4 right angles

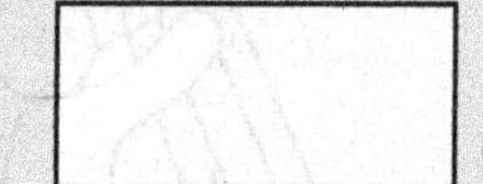
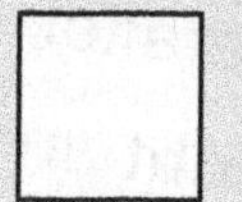

So . . . a square is both a rhombus and a rectangle according to its definition.

Square
2 pairs of opposite parallel sides
Lengths of opposite sides are equal
Lengths of all sides are equal
4 right angles

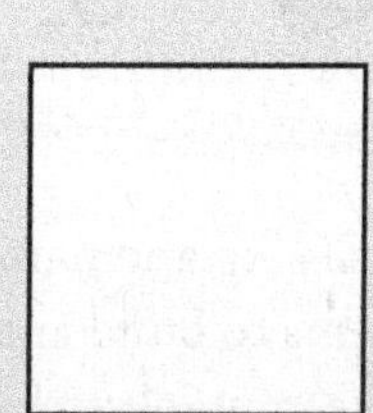

PRACTICE: Now you try

1. Use the word bank to list all words that describe the quadrilateral.	2. Draw the described quadrilateral and name it.	3. Circle the figure that is not a rectangle.
 ____ ____ ____ ____ ____ Polygon Quadrilateral Trapezoid Parallelogram Rhombus Rectangle Square	____ 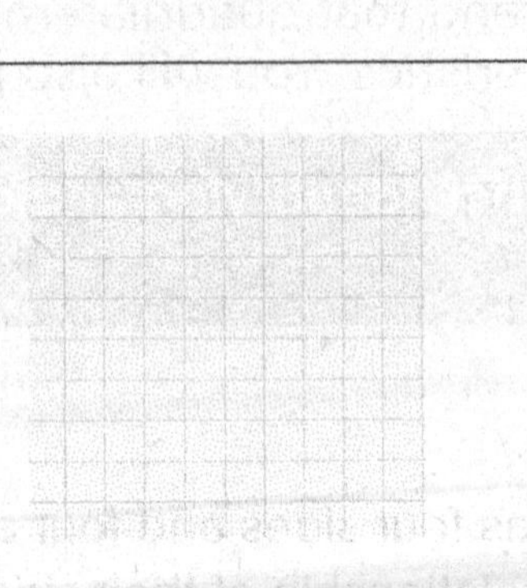 4 right angles, 4 equal sides	a. 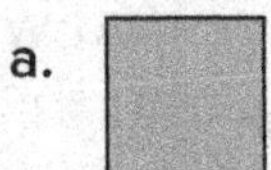 b.  c.  d. 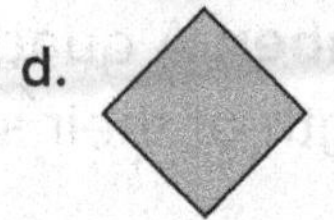

Sadi says that any square is also a rectangle. Corey says that any rectangle is also a square. Who is correct? Explain. Show your work and explain your thinking on a piece of paper.

ACE IT TIME!

	yes	no
Did you underline the question in the word problem?		
Did you circle the numbers or number words?		
Did you box the clue words that tell you what operation to use?		
Did you use a picture to show your thinking?		
Did you label your numbers and your picture?		
Did you explain your thinking and use math vocabulary words in your explanation?		

Math Vocabulary

quadrilateral
rectangle
square
equal
parallel
right angle

Math on the Move

Practice classifying and naming quadrilaterals at home or when out and about. Use household items to build and practice naming shapes. Some examples of items you can use are toothpicks, pretzel sticks, marshmallows, or even straws.

Partitioned Shapes, Equal Area, and Unit Fractions

FOLLOWING THE OBJECTIVE
You will partition shapes into parts with equal area and express those parts as unit fractions.

LEARN IT: Partition whole shapes into equal parts by making sure that the area of each part is the same or equal.

Separate a square into four equal parts, or draw lines to divide a whole into equal parts. You can use vertical, horizontal, or diagonal lines.

- A unit fraction represents each equal part.
- The equal parts have equal areas.
- When you partition figures into equal parts, you may make shapes with new names.

For each sample shown here, the unit fraction that represents each equal part of the whole is

$$\frac{1}{4}$$

Look at your denominator. There are **4** unit fraction ($\frac{1}{4}$) parts that make up the whole.

The area of each part is **4 square units**. The area of each part is equal.

Hint: You do not get perfect squares when partitioning the whole this way. But each of the triangles represents $\frac{1}{2}$ of a unit square. When two triangles are put together, it makes a unit square, so there are 4 squares in each of the equal parts of the whole. The area equals 4 square units.

$\frac{1}{4}$ $\frac{1}{4}$ $\frac{1}{4}$ $\frac{1}{4}$

Four equal parts

Four equal parts

Four equal parts

Four equal parts

PRACTICE: Now you try

Divide these shapes into equal parts (considering equal area) and name the unit fraction that each part makes.

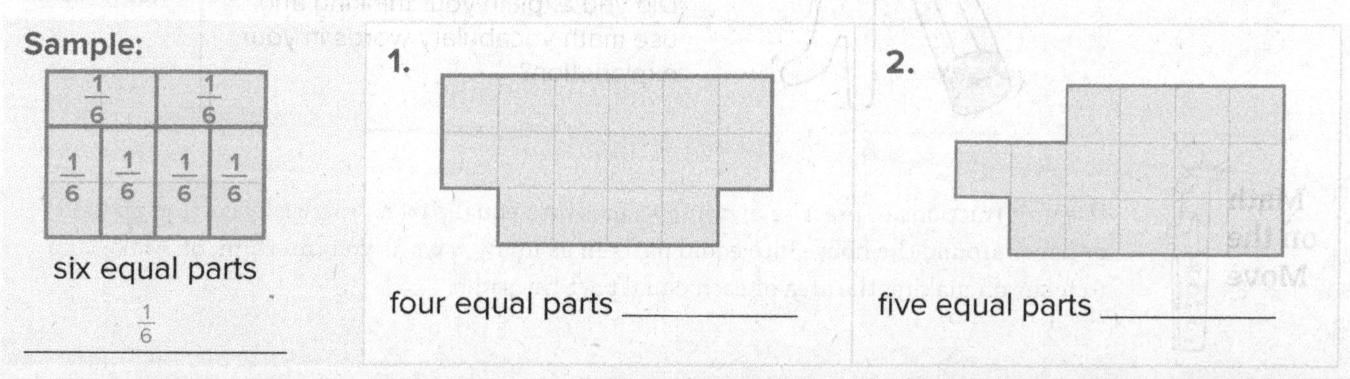

3. Show two ways to partition these rectangles into six equal parts.

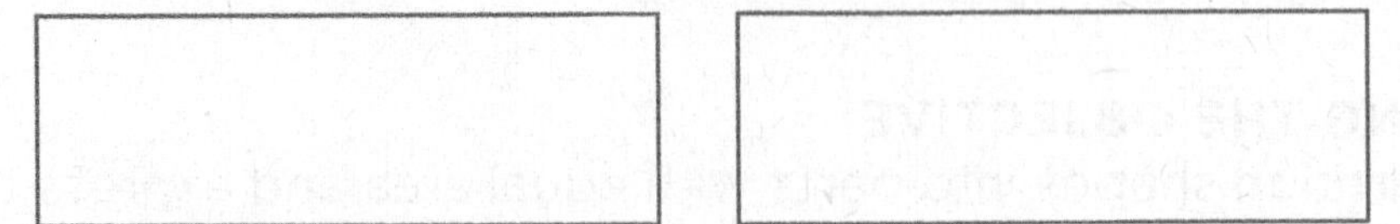

Write the unit fraction that represents each part. ____________

Divide the shapes into equal parts with equal area. Use the fraction given.

4. $\frac{1}{6}$

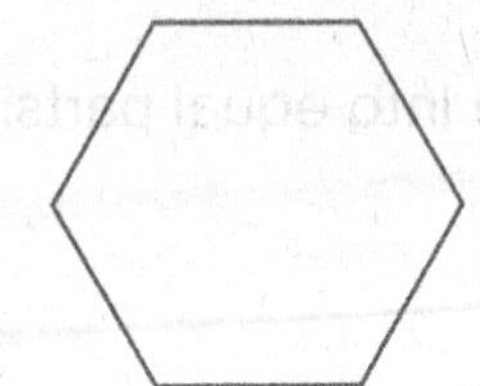

5. $\frac{1}{6}$

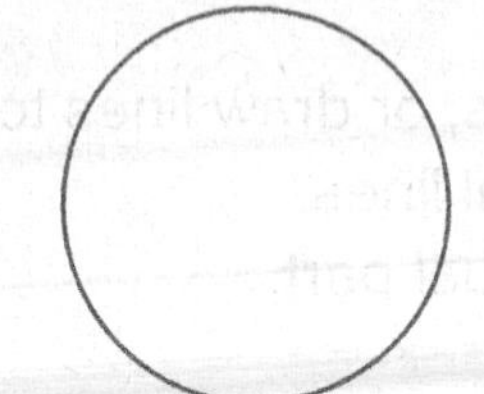

6. $\frac{1}{4}$

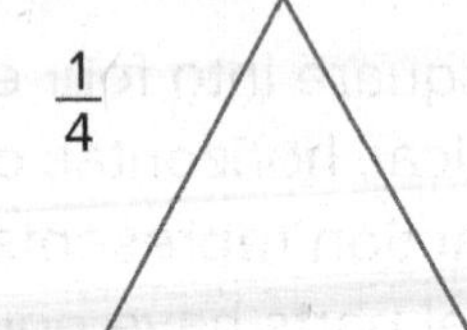

Marcia made 2 sandwiches, one for her and one for her sister. She tells her sister that she cut both sandwiches into fourths. Do you agree? (See pictures of how she cut the sandwiches below.) Show your work and explain your thinking on a piece of paper.

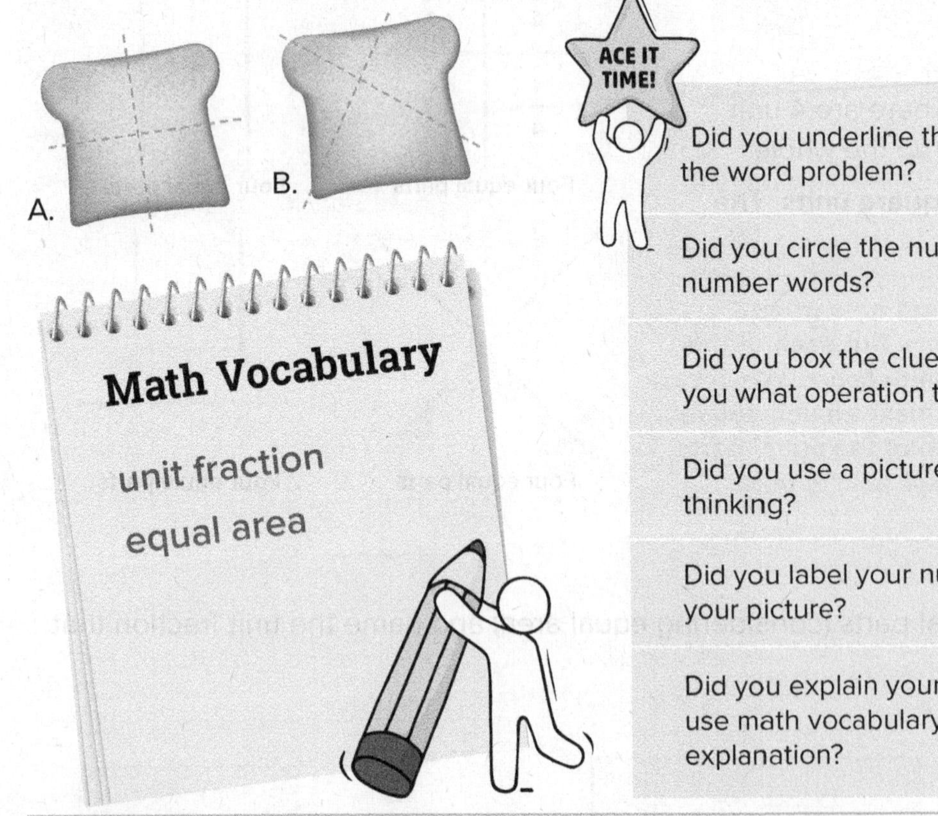

	yes	no
Did you underline the question in the word problem?		
Did you circle the numbers or number words?		
Did you box the clue words that tell you what operation to use?		
Did you use a picture to show your thinking?		
Did you label your numbers and your picture?		
Did you explain your thinking and use math vocabulary words in your explanation?		

Math on the Move

Use unit fractions to practice dividing shapes into equal parts. Practice by cutting up food or paper around the house into equal parts in as many ways as you can think of. Make sure to focus on making the area of each equal part the same.

REVIEW

Stop and think about what you have learned.

Congratulations! You've finished the lessons for this unit. This means you can identify, compare, and define two-dimensional shapes based on their attributes. You even practiced dividing shapes into equal parts with equal areas.

Now it's time to prove your skills with geometry. Solve the problems below! Use all of the methods you have learned.

Activity Section

Write whether the angle is acute, obtuse, or right.

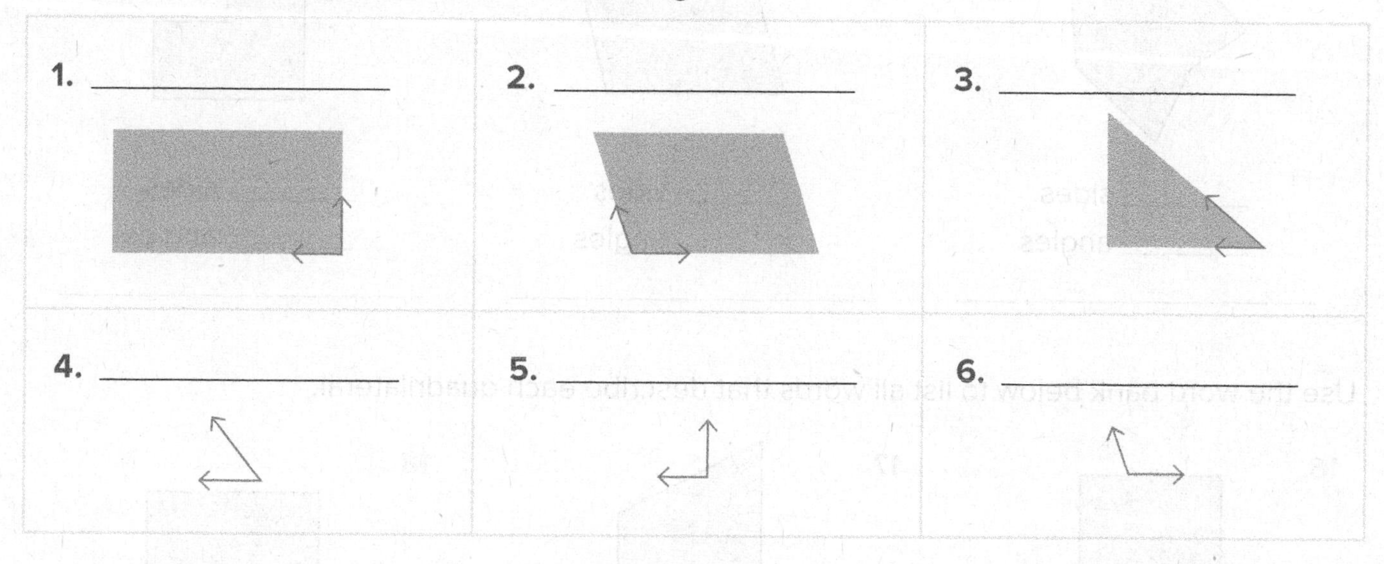

Write whether the highlighted sides in the polygons form parallel lines, intersecting lines, or perpendicular lines.

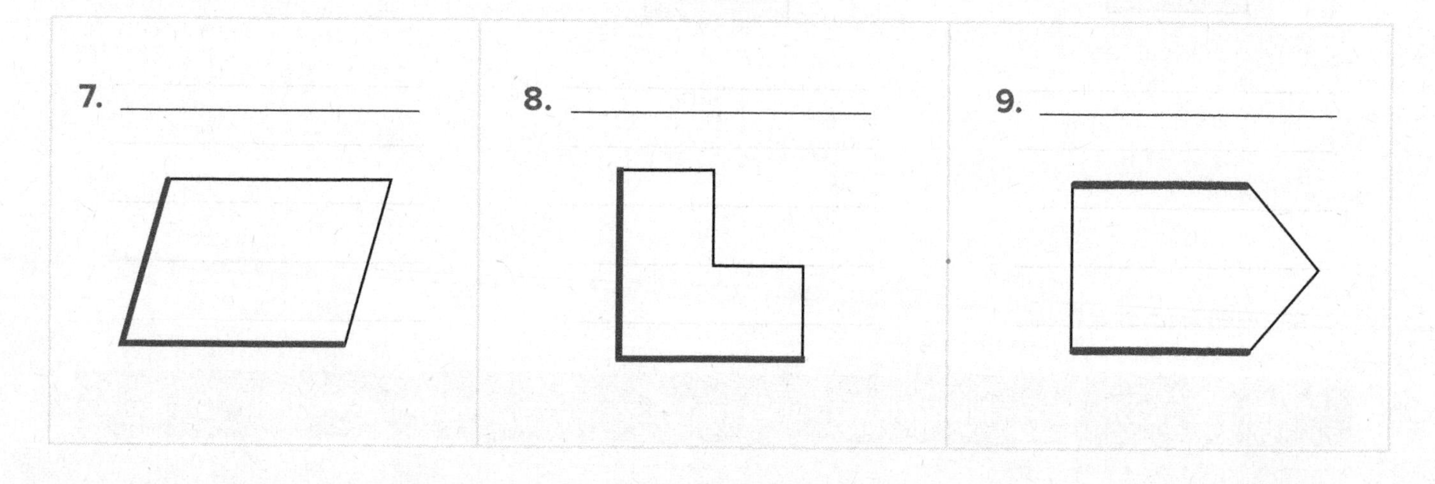

Write the number of sides and the number of angles. Name the polygon.

10.

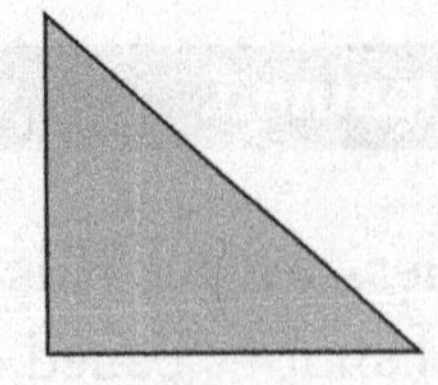

_______ sides
_______ angles
____

11.

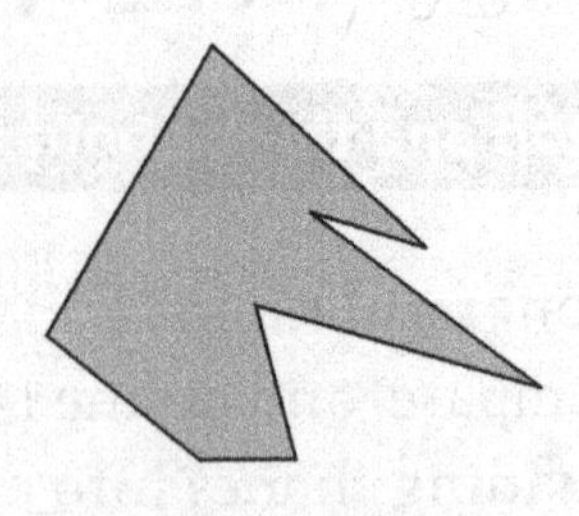

_______ sides
_______ angles
____

12.

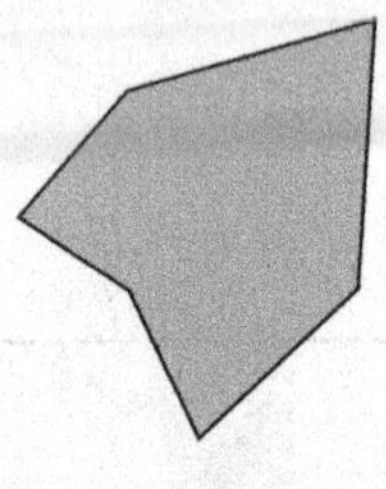

_______ sides
_______ angles
____

13.

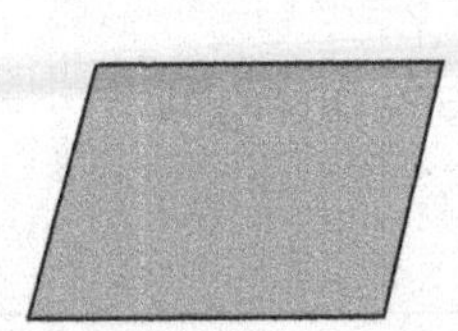

_______ sides
_______ angles
____

14.

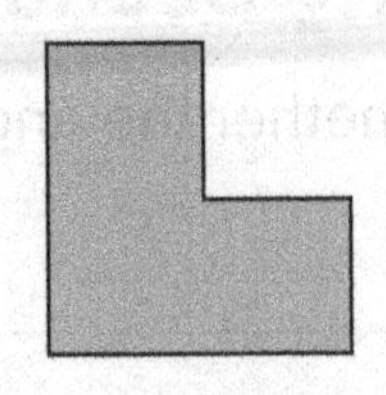

_______ sides
_______ angles
____

15.

_______ sides
_______ angles
____

Use the word bank below to list all words that describe each quadrilateral.

16.

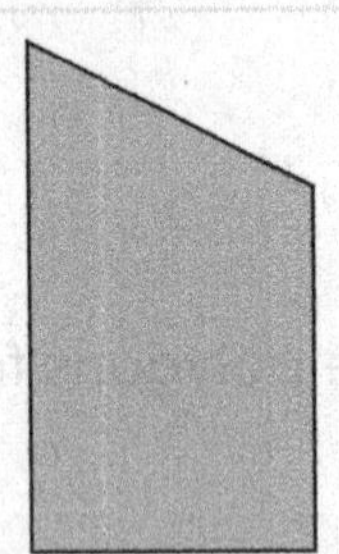

17.

18.

WORD BANK: Polygon Quadrilateral Trapezoid Parallelogram Rhombus Rectangle Square

Draw the described quadrilaterals and name them.

19.

4 sides
No parallel sides
No equal sides

____

20.

Exactly 1 pair of opposite parallel sides

2 right angles

____

21. Draw three quadrilaterals: a square, a rectangle, and a rhombus. Then draw a fourth quadrilateral that is not a square, a rectangle, or a rhombus.

Divide these shapes into equal parts (considering equal area) and name the unit fraction that each part makes.

22.

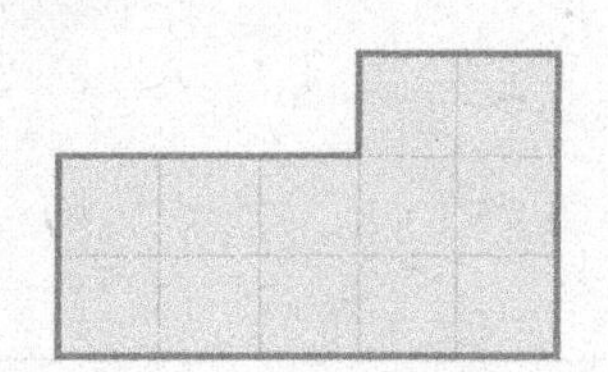

three equal parts

____

23.

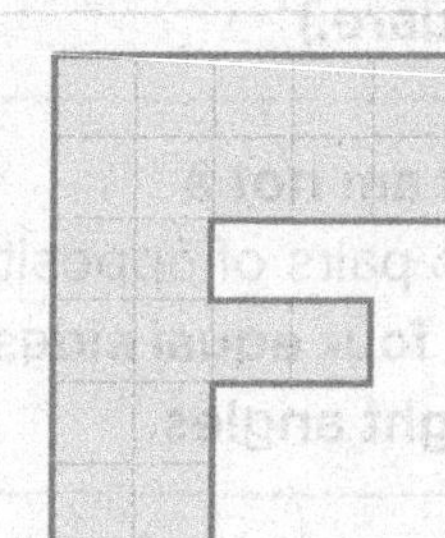

five equal parts

____

24. Show two ways to partition these rectangles into **eight** equal parts.

Write the unit fraction that represents each part.

____

25. Divide the shapes into equal parts with equal area. Use the fraction given.

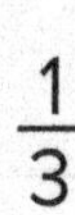

$\frac{1}{3}$

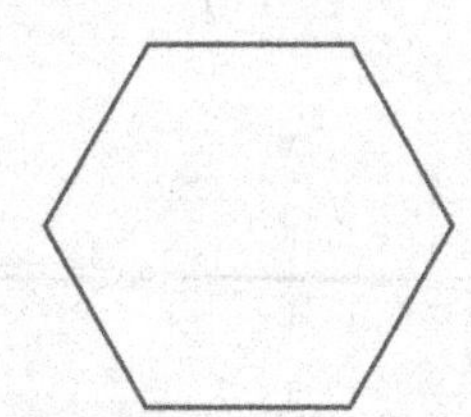

Understand the meaning of what you have learned and apply your knowledge.

You understand that shapes are defined by their attributes. You are also able to identify and draw examples of quadrilaterals.

Activity Section

Quadrilateral Riddles: Use the clues to identify the quadrilateral. Write the name of the shape and draw it. Write your own riddles for questions 5 and 6.

What shape am I?	Write the name.	Draw a picture of the shape.
1. I am a quadrilateral. I have exactly one pair of parallel sides.		
2. I am a parallelogram. I have two pairs of opposite parallel sides. I have four right angles. I have two pairs of equal sides. **(I am not a square.)**		
3. I am a quadrilateral. **I am not a trapezoid**. I have two pairs of opposite parallel sides. **I have four equal sides. I do not have four right angles.**		
4. I am a polygon. I have four sides. **None of my sides are parallel**. I have four angles.		
5.		
6.		

DISCOVER

Discover how you can apply the information you have learned.

You are able to divide shapes into parts with equal area and express each part as a unit fraction of the whole.

Activity Section

Mr. Williams makes lasagna for his children for dinner. He bakes the lasagna in a 6-by-8-inch casserole dish. He asks his daughter, Cassidy, to divide the casserole equally between herself and her three siblings.

1. How many square inches of lasagna did Mr. Williams make (area)?

____

2. What is the unit fraction that represents each sibling's equal portion of lasagna? How do you know?

____

3. What amount of lasagna (square inches) will each child get?

____

4. How do you know that each sibling's share is equal?

____

5. How did you determine the amount of lasagna each person will get? Is there another way that the lasagna could have been divided?

____

____

____

Answer Key

Unit 2: Addition and Subtraction Concepts

Patterns and Properties

Page 10 Practice: Now you try

1. 6 + 3; 9
2. 8 + 0; 8
3. Even
4. Even
5. Odd

Ace It Time: Allison is not correct because the sum of 2 odd numbers, 9 + 9, *always* equals an even number, 18.

Rounding to Ten

Page 12 Practice: Now you try

1. 20 (20-23-25-30)
2. 50 (40-45-48-50)

Ace It Time: The number is 73; it is the only number that will round to 70.

Rounding to 100

Page 14 Practice: Now you try

1. 200 (200-224-250-300)
2. 300

Ace It Time: Michael is correct. 673 is greater than 650 so it would round to 700.

Methods of Addition

Page 16 Practice: Now you try

1. 120
2. 989

Ace It Time: 175 + 175 + 175 = 525 minutes

Methods of Subtraction

Page 18 Practice: Now you try

1.
```
  700 + 80 + 8        242
– 500 + 40 + 6      + 546
  200 + 40 + 2        788
  = 242
```

2.
```
   5 15               1
   965              827
 – 138            + 138
   827              965
```

3.
```
  725 + 4             203
– 525 + 1           + 526
  200 + 3 = 203       729
```

Ace It Time: 153 – 76 = 77

Estimating Sums and Differences

Page 19 Practice: Now you try

1. 579 (250 + 330 = 580)
2. 216 (340 – 120 = 220)
3. 649 (270 + 380 = 650)
4. 191 (460 – 270 = 190)

Page 20

5. 724 (500 + 200 = 700)
6. 433 (700 – 200 = 500)

Ace It Time: 230 + 180 = 410 animals; rounding to the nearest ten provides the most reasonable amount of total adoptions.

Stop and Think! Unit 2 Review

Page 21 Activity Section 1: Patterns and Properties

1. 7 (0 + 7 =7)
2. 8 (7 + 1 = 8)
3. 3 (3 + 0 = 3)
4. 13 (3 + 10 = 13)
5. 5 (0 + 5 = 5)
6. 12 (4 + 8 = 12)
7. even
8. odd
9. even

Page 22 Activity Section 2: Rounding to 10 and 100

1. 60
2. 80
3. 50
4. 100
5. 600
6. 400
7. 200
8. 700

Page 23 Activity Section 3: Addition and Subtraction Methods

1.
```
  200 + 70 + 4
+ 100 + 10 + 2
  300 + 80 + 6 = 386
```

2.
```
   11
   573
 + 328
   901
```

3.
```
  675 + 3
+ 125 + 1
  800 + 4 = 804
```

4.
```
  200 + 70 + 4            162
– 100 + 10 + 2          + 112
  100 + 60 + 2 = 162      274
```

5.
```
   4 17              1
   857             519
 – 338            +338
   519             857
```

6.
```
  950 + 8           531
– 425 + 2         + 427
  525 + 6 = 531     958
```

Stop and Think! Unit 2 Understand

Page 24 Answers will vary but may include any 20 combinations of digits ranging from the least possible number 651 through the greatest number of 749. *Reminder:* Digits MUST NOT REPEAT in a number.

Examples could be: 749, 748, 746, 745, 743, 742, 741, 740, 739, 738, 651, 652, 653, 654, 657, 658, 659, 670, 671, 672; all round to 700. Students should realize there are many more than 20 possible numbers that will round to 700.

Stop and Think! Unit 2 Discover

Page 25

Mileage	Nearest 10	Nearest 100
437	440	400
289	290	300
83	80	100
Total 809	Total 810	Total 800

Page 26

1. Rounding to the nearest ten.
2. When you round to the nearest ten the actual number is closer to the ten than to the hundred.
3. 83 rounds to 100 because it is greater than the midpoint of 50.

Unit 3: Multiplication Concepts

Connecting Addition and Multiplication

Page 28 Practice: Now you try

1. 6 (3 +3)
2. 15 (5 + 5 + 5)
3. 24 (6 + 6 + 6 + 6)

Ace It Time:

2 (crabs) × 8 (legs) = 16 legs

5 (octopuses) × 8 (legs) = 40 legs

16 (crab legs) + 40 (octopus legs) = 56 legs

Since both crabs and octopuses have 8 legs, students could also add 5 (octopuses) + 2 (crabs) to equal 7 (animals) and then multiply by 8 (legs) = 56.

Skip-Counting

Page 29

1. 0, 3, 6, 9, 12; 4 × 3 = 12
2. 0, 4, 8, 12, 16; 4 × 4 = 16
3. 0, 5, 10, 15, 20; 4 × 5 = 20

Page 30 Practice: Now you try

1. 15 (5 skips of 3)
2. 18 (6 skips of 3)

Ace It Time: 8 (cars) × 4 (tires) is 8 jumps of 4 or 4 jumps of 8

8, 16, 24, 32 so 8 × 4 = 32 tires will be sold.

Multiplying with Arrays

Page 31 Practice: Now you try

1. 20 (4 rows of 5)
2. 27 (3 rows of 9)
3. 25 (5 rows of 5)
4. 6 (2 rows of 3)
5. 24 (6 rows of 4)
6. 42 (7 rows of 6)
7. 12 (3 rows of 4)
8. 10 (1 row of 10)
9. 32 (8 rows of 4)

Page 32

10. 2 × 4 = 8
11. 4 × 3 = 12
12. 4 × 6 = 24

Ace It Time: Adding equal groups should look like this:

Skip counting should look like this:

8, 16, 24, 32, 40, 48, or use a number line.

Arrays should show either 8 rows of 6 or 6 rows of 8, like this:

The Commutative Property

Page 33 Practice: Now you try

1. 4 × 7 = 28 and 7 × 4 = 28
2. 3 × 9 = 27 and 9 × 3 = 27
3. 4 × 2 = 8 and 2 × 4 = 8

Page 34

4. 12 (2 × 6 = 12)
5. 21 (7 × 3 = 21)
6. 45 (5 × 9 = 45)

Ace It Time: Liam could arrange his books either in:

1. 3 rows of 4 books
2. 4 rows of 3 books
3. 2 rows of 6 books
4. 6 rows of 2 books

Although 12 rows of 1 book is another combination since 1 row of 12 books didn't fit, this would not be a "reasonable" option because the book case only has 6 shelves.

Mastering Multiplication

Page 36 Practice: Now you try

1. 60
2. 90
3. 28
4. 72
5. 24
6. 15
7. 35
8. 6
9. 3

Ace It Time: Yessina and Olivia each have 6 markers: 2 (girls) × 6 (markers) = 12 markers

Kenji and Evita each have 9 markers: 2 (girls) × 9 (markers) = 18 markers; 12 + 18 = 30 markers in all

The Distributive Property

Page 37 Practice: Now you try

1. 36: 20 + 16 = 36

Page 38

2. 7 × 6 = 42

   Sums could include: 4 + 3, 5 + 2, 6 + 1.

Answers will be 2 of the following and will match the sums chosen:

(4 + 3) × 6
(4 × 6) + (3 × 6)
24 + 18 = 42

(5 + 2) × 6
(5 × 6) + (2 × 6)
30 + 12 = 42

(6 + 1) × 6
(6 × 6) + (1 × 6)
36 + 6 = 42

Ace It Time: No, Martha is not correct. (4 × 6) + (4 × 6) = 48 is not the correct product of 54. Students might also notice that the sum of 4 + 4 = 8, but the factor in the problem is 9.

The Associative Property

Page 39 Practice: Now you try

1. 9 × 4 = 3 × 12 = 36
2. 10 × 3 = 5 × 6 = 30
3. 6 × 2 = 3 × 4 = 12
4. 12 × 5 = 6 × 10 = 60
5. 20 × 1 = 4 × 5 = 20

Page 40

6. 10 × 5 = 50
7. (6 × 2) × 5
   12 × 5 = 60
8. 4 × (1 × 5)
   4 × 5 = 20
9. 3 × (2 × 6)
   3 × 12 = 36
10. (9 × 2) × 2
    18 × 2 = 36

Ace It Time: The number missing is 4. 7 × (2 × 4) = 7 × 8, which equals 56.

Number Patterns in Multiplication

Page 41 Learn It:

Green Crayon

a) All the digits in the ones place are even and repeat 2, 4, 6, 8, 0 . . . .

b) All the digits in the ones place are even and repeat 4, 8, 2, 6, 0 . . . .

c) The products in column 4 are twice as large as (×2) the products in column 2.

Blue Crayon

a) The digits in the ones place alternate between odd and even, and repeat 0, 5, 0, 5 . . . .

b) The digits in the ones place are all zeros.

c) The products in row 10 are twice as large as (×2) the products in row 5.

Yellow Crayon

a) The digits in the ones place alternate between odd and even.

b) The digits in the ones place are all even.

c) The products in column 6 are twice as large as (×2) the products in column 3.

Red Crayon

a) The products alternate between odd and even.

b) The products are all even.

c) No, there is not a row that has only odd numbers.

Page 42 Practice: Now you try

1. 81 (odd × odd = odd)
2. 12 (even × odd =even)
3. 20 (even × even = even)
4. 28 (odd × even = even)
5. Alike—adding or multiplying two even numbers always equals an even number.

   Different—Adding two odd numbers equals an even sum, while multiplying two odd numbers equals an odd product. Adding an even number and an odd number equals an odd sum, while multiplying an even number and an odd number equals an even product.

Ace It Time: Jake's two mistakes are: 27 should be 28 (7 × 4 = 28), and 64 should be 63 (7 × 9 = 63).

Multiplying with Tens

Page 44 Practice: Now you try

1. 3 (tens) × 6
   18 (tens)
   180
2. 5 (tens) × 9
   45 (tens)
   450
3. 6 (tens) × 4
   24 (tens)
   240

Ace It Time: 70 (pages) × 5 (notebooks) = 350 pages in all

Stop and Think! Unit 3 Review

Page 45 Activity Section

1. 36
2. 6: 2 + 2 + 2
3. 12: 4 + 4 + 4
4. 25, 35, 45, 55
5. 10, 40, 50, 60
6. 100, 200, 500, 600
7. 14, 16, 18, 20
8. 40, 44, 48, 52
9. 3, 12, 15, 18

Page 46

10. 28: 7 × 4 = 28
11. 54: 9 × 6 = 54
12. 15: 3 × 5 = 15
13. 56: 7 × 8 = 56
14. 20
15. 42: break it up as 3 + 3 or 4 + 2; arrays will reflect the addition fact selected

Page 47

16. (3 × 4) × 3
    12 × 3
    36
17. 5 × (5 × 2)
    5 × 10
    50
18. 7 × (4 × 2)
    7 × 8
    56
19. 150
20. 120
21. 160
22. 450
23. 20
24. 40
25. 60
26. 14
27. 42

28. 45
29. 10
30. 0
31. 48
32. 15
33. 0
34. 24

Stop and Think! Unit 3 Understand

Page 48

Ways to solve 8 × 9 using the distributive property include:

1. (1 + 7) × 9
   1 × 9 + 7 × 9
   9 + 63
   72
2. (2 + 6) × 9
   2 × 9 + 6 × 9
   18 + 54
   72
3. (3 + 5) × 9
   3 × 9 + 5 × 9
   27 + 45
   72
4. (4 + 4) × 9
   4 × 9 + 4 × 9
   36 + 36
   72
5. 8 × (1 + 8)
   8 × 1 + 8 × 8
   8 + 64
   72
6. 8 × (2 + 7)
   8 × 2 + 8 × 7
   16 + 56
   72

Bonus: Yes, there are other ways in addition to the ones above: 8 × (3 + 6); 8 × (4 + 5)

Stop and Think! Unit 3 Discover

Page 50 Activity Section

1. Multiply × 4; missing numbers include 16, 20.
2. Answers will vary, but should be about the number of pencils and packages. Sample example: One package of pencils has 4 pencils. How many pencils are in 5 packages?
3.

Mealworms	1	2	3	4	5
Legs	6	12	18	24	30

4. 30 legs

Unit 4: Division Concepts

Making Equal Groups

Page 52 Practice: Now you try

1. 8 triangles divided into 4 equal groups = 2 triangles in each group

Ace It Time: 28 marbles divided into 4 bags equally = 7 marbles in each bag

Connecting Subtraction and Division

Page 54 Practice: Now you try

1. 3:
   15 − 5 = 10 → 10 − 5 = 5 → 5 − 5 = 0
2. 3:
   21 − 7 = 14 → 14 − 7 = 7 → 7 − 7 = 0
3. 4:
   8 − 2 = 6 → 6 − 2 = 4 → 4 − 2 = 2 → 2 − 2 = 0
4. 3:
   30 − 10 = 20 → 20 − 10 = 10 → 10 − 10 = 0

Ace It Time: 36 (books) ÷ 4 (shelves) = 9 (number of books on each shelf)

Dividing with Arrays

Page 55 Practice: Now you try

1. 6; should reflect 3 rows of 6

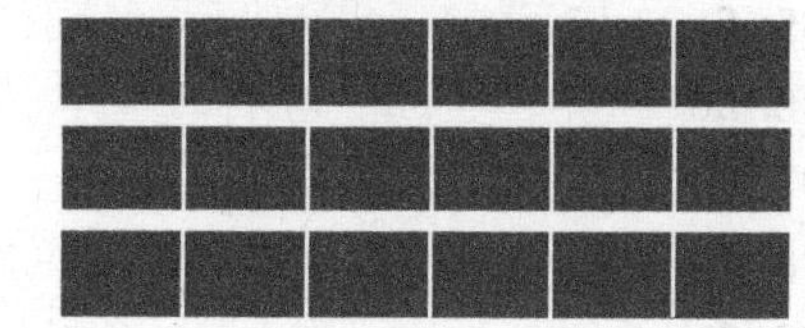

2. 3; should reflect 4 rows of 3

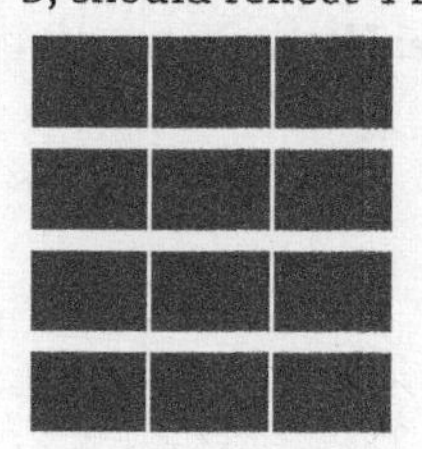

Page 56

3. 5; should reflect 5 rows of 5

Ace It Time: 6 cards in each row × 6 rows would equal 36 cards.

Connecting Multiplication and Division

Page 57 Practice: Now you try

1. 6; 5 × 6= 30; 30 ÷ 5 = 6
2. 9; 4 × 9 = 36; 36 ÷ 4 = 9

Page 58

3. 5 × 10 = 50; 10 × 5 = 50; 50 ÷ 10 = 5; 50 ÷ 5 = 10
4. 6 × 3 = 18; 3 × 6 = 18; 18 ÷ 6 = 3; 18 ÷ 3 = 6

Ace It Time: Marian rode the Ferris wheel 6 times; 24 tickets total ÷ 4 tickets per ride = 6 rides.

Multiplying and Dividing by 0 and 1

Page 60 Practice: Now you try

1. 9
2. 0
3. 0

Ace It Time: Mary is not correct. Mary has 6 vases but no flowers. Jane also has 6 vases but she has 1 flower in each vase, so she has 6 flowers, which is not the same as no flowers.

Division Facts

Page 62 Practice: Now you try

1. 8 ÷ 2 = 4 and 8 ÷ 4 = 2

Ace It Time: There are 9 chapters in Caylie's book; 72 pages ÷ 8 pages per chapter = 9 chapters.

Find the Unknown Number

Page 64 Practice: Now you try

1. 28
2. 4
3. 4
4. 12
5. 5
6. 6
7. 54
8. 6
9. 3 × 4 = 12; 12 ÷ 3 = 4; unknown number = 4
10. 7 × 7 = 49; 49 ÷ 7 = 7; unknown number = 7

Ace It Time: 35 cookies total; equation: ______ ÷ 5 = 7; 5 × 7 = ______ (answer = 35)

Two-Step Problems and Equations

Pages 65–66 Practice: Now you try

1. How many total baseball cards does he have? 48 + 12 = 60

   How many pages will he fill? 60 ÷ 12 = 5

Ace It Time: How many total players entered the games? 6 × 7 = 42

How many ice pops are left over? 100 – 42 = 58

Stop and Think! Unit 4 Review

Page 67 Activity Section

1. 15 ÷ 3= 5
2. 12 ÷ 4 = 3
3. 16 ÷ 8 = 2

Page 68

4. 50, 40, 30, 20
5. 12, 10, 8, 6
6. 45, 27, 18, 9
7. 35, 28, 21, 14
8. 95, 85, 70, 65
9. 16, 12, 8, 4
10. 3 × 10 = 30
    10 × 3 = 30
    30 ÷ 10 = 3
    30 ÷ 3 = 10
11. 6 × 8 = 48
    8 × 6 = 48
    48 ÷ 6 = 8
    48 ÷ 8 = 6
12. 6 × 7 = 42
    7 × 6 = 42
    42 ÷ 7 = 6
    42 ÷ 6 = 7
13. 18 ÷ 3 = 6
    18 ÷ 6 = 3
    6 × 3 = 18
    3 × 6 = 18
14. 5 × 8 = 40
    8 × 5 = 40
    40 ÷ 8 = 5
    40 ÷ 5 = 8
15. 54 ÷ 6 = 9
    54 ÷ 9 = 6
    9 × 6 = 54
    6 × 9 = 54
16. 8 × 8 = 64
    64 ÷ 8 = 8

Page 69

17. 8
18. 6
19. 9
20. 7
21. 4
22. 4
23. 5
24. 6
25. 5
26. 24
27. 18
28. 7
29. 7
30. 3
31. 53
32. 2
33. 8
34. 23
35. 6
36. 56
37. 6
38. 7
39. 3
40. 7

Stop and Think! Unit 4 Understand

Page 70 Activity Section

These are possible ways you can divide 24 tiles (in addition to the given way of 4 rows of 6 tiles)

1. 3 rows of 8 tiles or 8 rows of 3 tiles: 24 ÷ 3 = 8; 24 ÷ 8 = 3; 3 × 8 = 24; 8 × 3 = 24
2. 2 rows of 12 tiles or 12 rows of 2 tiles: 24 ÷ 2 = 12; 24 ÷ 12 = 2; 2 × 12 = 24; 12 × 2 = 24
3. 1 row of 24 tiles or 24 rows of 1 tile: 24 ÷ 1 = 24; 24 ÷ 24 = 1; 1 × 24 = 24; 24 × 1 = 24

Stop and Think! Unit 4 Discover

Page 71 Activity Section

1. 60 (money raised) ÷ 5 (amount per tub) = 12 (tubs of cookie dough sold)

Page 72

2. 20 (money given) ÷ 3 (amount per tub) = 6 (tubs of cookie dough bought). She can't buy 7 tubs because she doesn't have enough money.
3. Justin raised $24.00 ÷ 4 (amount per tub) = 6 tubs of cookie dough

   Bryce raised $24.00 ÷ 3 (amount per tub) = 8 tubs of cookie dough

   Even though they raised the same amount of money, Bryce sold more cookie dough.

Unit 5: Fraction Concepts

Understanding Fractions

Page 74 Practice: Now you try

1. not equal
2. not equal
3. equal

Page 75

4. thirds
5. halves
6. sixths

Ace It Time: No, the sizes of the wholes are different.

Unit Fractions and Other Fractions

Page 77 Practice: Now you try

1. $\frac{1}{8}$
2. $\frac{1}{4}$
3. $\frac{1}{5}$

4. The student should divide each rectangle into fourths. Possible answers include:

5. $\frac{3}{4}$

6. $\frac{2}{4}$

7. $\frac{4}{8}$

Page 78

8. Shade 4 parts.

9. Shade 5 parts.

10. Shade 3 parts.

Ace It Time: 2 more pieces

Fractions on a Number Line

Page 80 Practice: Now you try

1. $\frac{2}{6}, \frac{4}{6}, \frac{5}{6}$

Ace It Time: Girl

Relate Fractions and Whole Numbers

Page 82 Practice: Now you try

1. $\frac{6}{6}$, 1

2. $\frac{9}{3}$, 3

Ace It Time: Lloyd ate more pita chips. He ate 6 and Tammy ate 3. $\frac{9}{3} = 3$

Compare Fractions with the Same Denominator

Page 84 Practice: Now you try

1. $\frac{4}{6} > \frac{1}{6}, \frac{1}{6} < \frac{4}{6}$

2. $\frac{2}{6} < \frac{5}{6}, \frac{5}{6} > \frac{2}{6}$

Ace It Time: Charlene read more pages. $\frac{5}{8} > \frac{2}{8}$

Compare Fractions with the Same Numerator

Page 86 Practice: Now you try

1. $\frac{2}{3} > \frac{2}{6}, \frac{2}{6} < \frac{2}{3}$

2. $\frac{1}{6} < \frac{1}{3}, \frac{1}{3} > \frac{1}{6}$

Ace It Time: Pam was closer to being done.

Comparing and Ordering Fractions

Page 88 Practice: Now you try

1. $\frac{1}{8}, \frac{1}{3}, \frac{1}{2}$

2. $\frac{2}{6}, \frac{4}{6}, \frac{6}{6}$

3. $\frac{1}{8}, \frac{2}{8}, \frac{6}{8}$

4. $\frac{1}{4}, \frac{2}{4}, \frac{3}{4}$

5. $\frac{2}{6}, \frac{2}{5}, \frac{2}{3}$

6. $\frac{3}{6}, \frac{3}{4}, \frac{3}{3}$

Ace It Time: Wednesday

Equivalent Fractions

Page 90 Practice: Now you try

1. (Row 1) $\frac{1}{4}, \frac{4}{4}$; (Row 2) $\frac{1}{8}, \frac{4}{8}, \frac{5}{8}, \frac{7}{8}$; circle equivalent fractions: $\frac{1}{4}$ and $\frac{2}{8}$, $\frac{2}{4}$ and $\frac{4}{8}$, $\frac{3}{4}$ and $\frac{6}{8}$

Ace It Time: Yes, because $\frac{2}{3} = \frac{4}{6}$

Stop and Think! Unit 5 Review

Page 91 Activity Section

1. $\frac{3}{4}$: Student should divide each circle into 4 equal parts.

2. $1\frac{1}{3}$: Student should divide at least 1 granola bar into 3 equal parts.

3. $\frac{2}{4}$

4. $\frac{2}{6}$

5. $\frac{1}{2}$

6. $\frac{3}{4}$

Page 92

7. Shade 2 rectangles.

8. Shade 1 triangle.

9. Shade 1 triangle.

10. Shade 3 wedges.

11. $\frac{1}{4}, \frac{3}{4}, \frac{4}{4}$

12. $\frac{1}{6}, \frac{2}{6}, \frac{4}{6}, \frac{5}{6}, \frac{6}{6}$ (or 1)

13. $\frac{7}{8}$

14. Point B

15. $\frac{2}{8}$

Page 93

16. 1

17. $\frac{8}{2}$ or $\frac{12}{3}$ or $\frac{16}{4}$ . . .

18. 2

19. $\frac{6}{2}$ or $\frac{9}{3}$ or $\frac{12}{4}$ . . .

20. $\frac{1}{1}$ or $\frac{2}{2}$ or $\frac{3}{3}$ or $\frac{4}{4}$ . . .

21. 3

22. $\frac{3}{8} < \frac{3}{6}$

23. $\frac{2}{4} > \frac{1}{4}$

24. $\frac{1}{3} < \frac{3}{3}$

25. $\frac{2}{4} > \frac{2}{6}$

26. $\frac{5}{8} > \frac{2}{8}$

27. $\frac{1}{6} < \frac{1}{3}$

28. $\frac{5}{6} > \frac{2}{6}$

29. $\frac{4}{4} > \frac{4}{6}$

30. $\frac{1}{2} < \frac{2}{2}$

31. $\frac{5}{6} > \frac{5}{8}$

32. $\frac{2}{6} < \frac{2}{3}$

33. $\frac{2}{4} < \frac{3}{4}$

34. $\frac{1}{4}, \frac{1}{3}, \frac{1}{2}$

35. $\frac{2}{8}, \frac{5}{8}, \frac{6}{8}$

36. $\frac{1}{6}, \frac{2}{6}, \frac{6}{6}$

37. $\frac{1}{2}, \frac{1}{6}, \frac{1}{8}$

38. $\frac{7}{8}, \frac{3}{8}, \frac{1}{8}$

39. $\frac{4}{4}, \frac{3}{4}, \frac{1}{4}$

40. $\frac{4}{8}$: Divide one model into 4 equal parts and shade 2 parts. Divide the other model into 8 equal parts and shade 4 parts.

41. $\frac{4}{6}$: Divide one model into 3 equal parts and shade 2 parts. Divide the other model into 6 equal parts and shade 4 parts.

42. $\frac{3}{3}$: Divide one model into 4 equal parts and shade 4 parts. Divide the other model into 3 equal parts and shade 3 parts.

Stop and Think! Unit 5 Understand

Page 94 Activity Section

She is right because $\frac{3}{4}$ is equal to $\frac{6}{8}$.

Stop and Think! Unit 5 Discover

Page 95 Activity Section

Melissa had the longest pencil; 9 inches = $\frac{3}{4}$ of a foot.

Unit 6: Data Concepts

Picture Graphs and Bar Graphs

Page 97 Practice: Now you try

1. Picture graph: Answers will vary depending on the key created.

Bar graph: Bars should be drawn and shaded to correspond to Reading—12 students, Math—14 students, Writing—8 students and Science—10 students.

2. 44
3. 6
4. 24
5. 4

Page 98

Ace It Time: Reading (12) + Math (14) = 26 students; Science (10) + Writing (8) = 18 students; 26 – 18 = 8 more students

Line Plots

Page 100 Practice: Now you try

1.

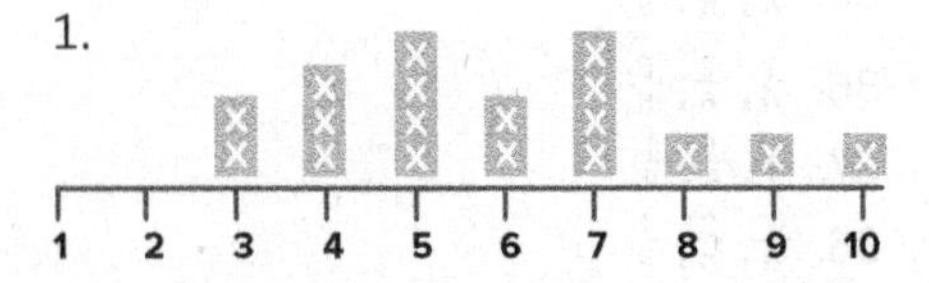

2. 18
3. 2
4. 9
5. 2

Page 101

Ace It Time: 4 (7) + 1 (8) + 1 (9) + 1 (10) = 7 students; 3 (4) + 2 (3) + 0 (1 & 2) = 5; 7 – 5 = 2 more students

Stop and Think! Unit 6 Review

Page 102 Activity Section

1. 16
2. 8
3. 20

Page 103

4. 5
5. 4
6. 7
7. Samantha, Brandon, Tyler, Megan, Noah

Page 104

8. Bar graph:

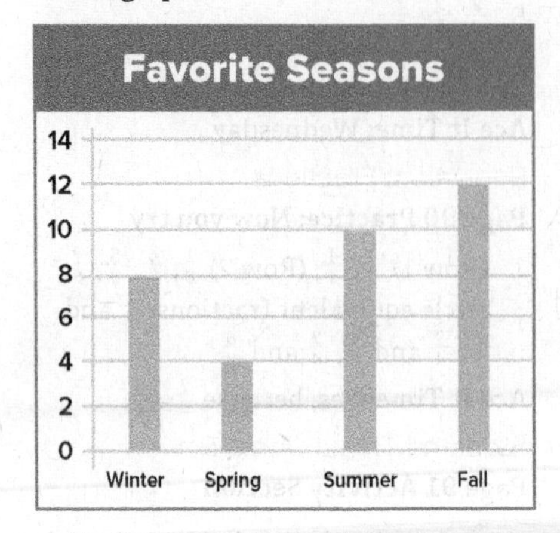

Picture graph answers will vary based on key selected but must reflect the given data on Favorite Seasons.

9. Line plot:

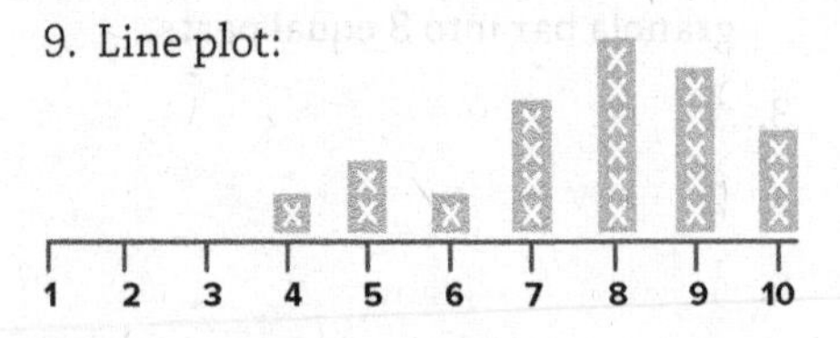

10. & 11. Questions written by students will vary.

Stop and Think! Unit 6 Understand

Page 105 Activity Section

Student's graph must accurately reflect the data given in both vertical and horizontal formats.

Vertical Format

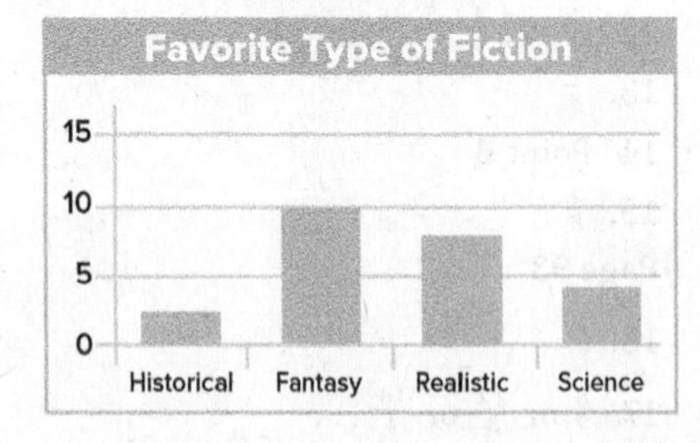

Horizontal Format

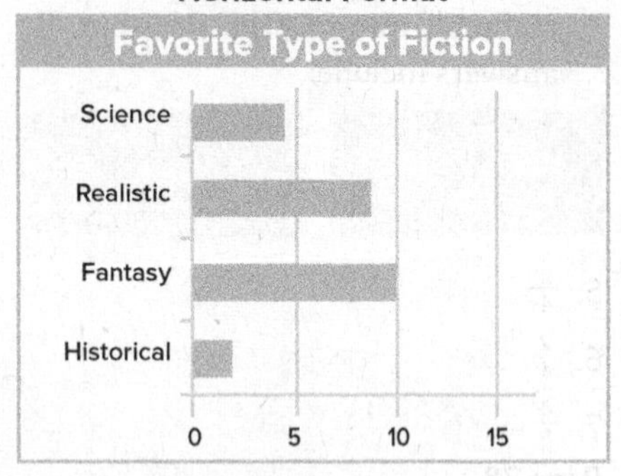

Stop and Think! Unit 6 Discover

Page 106 Activity Section

1. Top bar—Biography
   2nd bar—Poetry
   3rd bar—Nonfiction
   4th bar—Fiction
2. Answers may vary but might resemble the following:
   a. Fiction = 2 × 6 (Biography) = 12
   b. Poetry = 12 (Fiction) – 2 = 10
   c. Nonfiction = 2 × 12 (Fiction) = 24
   d. Biography—had the smallest bar and was half of Fiction: 12 ÷ 2 = 6
3. The star should be placed on the graph a little more than half way between 15 and 20. Explanations will vary but should indicate understanding of 18's place between 15 and 20.

Unit 7: Measurement Concepts

Telling and Writing Time

Page 108 Practice: Now you try

1. 8:07
2. 4:49
3. 6:40

Ace It Time: 3:41

Intervals of Time

Page 111 Practice: Now you try

1. Start time, 6:50 p.m.; Elapsed time, 45 minutes; End time, 7:35 p.m.
2. Start time, 6:10 p.m.; Elapsed time, 48 minutes; End time, 6:58 p.m.

Ace It Time: 3:20 p.m. Robert was at the library for a total of 1 hour and 10 minutes, so he had to get there at 3:20 p.m.

Length, Mass, and Liquid Volume

Page 112 Practice: Now you try

1. $3\frac{1}{2}$ in.

Page 113

2. gram
3. more than 1 liter

Ace It Time: 84 grams ÷ 12 cookies = 7 grams per cookie

Stop and Think! Unit 7 Review

Page 114 Activity Section

1. 8:18
2. 4:25
3. 6:50
4. The short hand should be slightly past the 7. The long hand should be pointing at the 1st tick mark past the 2.
5. The short hand should be pointing between the 2 and the 3. The long hand should be pointing at the 3rd tick mark past the 4.
6. The short hand should be pointing slightly before the 6. The long hand should be pointing at the 3rd tick mark past the 9.

Page 115

7. Start time, 12:45 p.m.; Elapsed time, 2 hours 10 minutes; End time, 2:55 p.m.
8. Start time, 12:20 p.m.; Elapsed time, 2 hours 40 minutes; End time, 3:00 p.m.
9. Start time, 2:15 p.m.; Elapsed time, 40 minutes; End time, 2:55 p.m.
10. Start time, 5:50 p.m.; Elapsed time, 1 hour 35 minutes; End time, 7:25 p.m.
11. Start time, 3:45 p.m.; Elapsed time, 1 hour 15 minutes; End time, 5:00 p.m.
12. Start time, 12:35 p.m.; Elapsed time, 2 hours 45 minutes; End time, 3:20 p.m.

Page 116

13. $4\frac{3}{4}$ inches
14. $2\frac{1}{2}$ inches
15. $3\frac{1}{4}$ inches
16. 4 inches
17. gram
18. kilogram
19. gram
20. kilogram
21. kilogram
22. gram
23. about 1 liter
24. more than 1 liter
25. less than 1 liter

Stop and Think! Unit 7 Understand

Page 117 Activity Section

Answers will vary.

Stop and Think! Unit 7 Discover

Page 118 Activity Section

45 minutes + 25 minutes + 20 minutes = 1 hour 30 minutes, so Doug and his brother should leave their home at 11:30 p.m.

Unit 8: Perimeter and Area Concepts

What Is Perimeter?

Page 120 Practice: Now you try

1. 30 cm
2. 22 cm
3. 21 cm

Ace It Time: $100.00

Using Perimeter to Find Missing Sides

Page 121 Practice: Now you try

1. n = 12 cm
2. n = 12 in.

Page 122

3. n = 39 in.
4. n = 14 cm
5. n = 17 cm

Ace It Time: 36 inches ÷ 4 equal sides = 9 inches per side

What Is Area?

Page 123 Practice: Now you try

1. 12 square units
2. 21 square units

Page 124

3. Answers will vary.
4. 8 square units
5. 15 square units

Ace It Time: Chris is correct.

More Work with Area

Page 125 Practice: Now you try

1. 18 square feet
2. 36 square inches
3. 10 square inches

Page 126

4. 81 square meters
5. 35 square feet

Ace It Time: 8 feet

Use Smaller Rectangles to Find Area

Page 128 Practice: Now you try

1. (2 × 6) + (2 × 3), 12 + 6 = 18 square meters
2. 8 × 4 = 32, 4 × 4 = 16, 32 + 16 = 48 square cm

Ace It Time: Sheila is wrong. The area is 132 square meters.

Relating Perimeter and Area

Page 130 Practice: Now you try

1. Area = 12 square feet; Perimeter = 14 feet.
2. Answers may vary, but the area of the rectangle should equal 12 square feet and the perimeter should not equal 14 ft.
3. Area = 36 square inches; Perimeter = 26 inches.
4. Answers may vary, but the area of the rectangle should equal 36 square feet and the perimeter should not equal 26 feet.

Ace It Time: Today, because the area of Mrs. Steel's garden is greater.

Stop and Think! Unit 8 Review

Page 131 Activity Section

1. 72 feet
2. 74 yards
3. 38 centimeters

4. 6 feet
5. 17 centimeters
6. 9 inches
7. 54 square centimeters
8. 72 square feet

Page 132

9. (3 × 6) + (3 × 3), 18 + 9 = 27 square meters
10. (7 × 5) + (7 × 6), 35 + 42 = 77 square inches
11. 56 square units
12. 63 square units
13. 20 × 5 = 100; 11 × 8 = 88; 100 + 88 = 188 square centimeters
14. 8 × 3 = 24; 9 × 3 = 27; 24 + 27 = 51 square inches

Page 133

15. Area = 15 square inches; Rectangle 1 (1 by 7 rectangle) – area = 7 square inches; Rectangle 2 (2 by 6 rectangle) – area = 12 square inches; Rectangle 3 (4 by 4 rectangle) – area = 16 square inches
16. Area = 12 square feet; Perimeter = 16 feet; Rectangle 1 (1 by 12 rectangle) – perimeter = 26 feet; Rectangle 2 (3 by 4 rectangle) – perimeter = 14 feet

Stop and Think! Unit 8 Understand

Page 134 Activity Section

He needs to buy 8 feet of fence.

Stop and Think! Unit 8 Discover

Page 135 Activity Section

In any order:

Table set-up #1: 1 by 5 rectangle (5 tables – area = 5 square units)

Table set-up #2: 2 by 4 rectangle (8 tables – area = 8 square units)

Table set-up #3: 3 by 3 rectangle (9 tables – area = 9 square units)

Unit 9: Geometry Concepts

Shapes and Attributes

Page 137 Practice: Now you try

1. acute
2. obtuse
3. right
4. perpendicular lines
5. parallel lines
6. intersecting lines

Page 138

7. 5 sides, 5 angles
8. 4 sides, 4 angles
9. 8 sides, 8 angles
10. 6 sides, 6 angles
11. 6 sides, 6 angles
12. 3 sides, 3 angles

Ace It Time: Four yards per side × 4 sides = 16 yards × \$2 per yard = \$32

Quadrilaterals

Page 140 Practice: Now you try

1. Polygon; Quadrilateral; Trapezoid
2. Square
3. B

Ace It Time: Sadi is correct because the opposite sides of a square are equal. A square is a special type of rectangle. A rectangle is any shape with four sides and four right angles. All squares are rectangles, but not all rectangles are squares.

Partitioned Shapes, Equal Area, and Unit Fractions

Page 141 Practice: Now you try

1. $\frac{1}{4}$
2. $\frac{1}{5}$

Page 142

3. $\frac{1}{6}$
4. The shape should be divided into 6 equal parts.
5. The shape should be divided into 6 equal parts.
6. The shape should be divided into 4 equal parts.

Ace It Time: Disagree: Sandwich B is not cut into 4 equal parts, so it cannot be fourths.

Stop and Think! Unit 9 Review

Page 143 Activity Section

1. right angle
2. obtuse angle
3. acute angle
4. acute angle
5. right angle
6. obtuse angle
7. intersecting lines
8. perpendicular lines
9. parallel lines

Page 144

10. 5 sides, 5 angles, pentagon
11. 3 sides, 3 angles, triangle
12. 8 sides, 8 angles, octagon
13. 6 sides, 6 angles, hexagon
14. 4 sides, 4 angles, quadrilateral
15. 6 sides, 6 angles, hexagon
16. Polygon, Quadrilateral, Parallelogram, Rectangle
17. Polygon, Quadrilateral, Trapezoid
18. Polygon, Quadrilateral, Parallelogram, Rhombus, Rectangle, Square

Page 145

19. quadrilateral
20. trapezoid
21. Fourth quadrilateral could be a trapezoid or another quadrilateral that is not a square, rectangle, or rhombus.
22. $\frac{1}{3}$
23. $\frac{1}{5}$
24. $\frac{1}{8}$
25. The shape should be divided into 3 equal parts.

Stop and Think! Unit 9 Understand

Page 146 Activity Section

1. trapezoid
2. rectangle
3. rhombus
4. quadrilateral
5. Answers will vary.
6. Answers will vary.

Stop and Think! Unit 9 Discover

Page 147 Activity Section

1. 48 square inches
2. $\frac{1}{4}$ because there are 4 siblings total.
3. 12 square inches
4. Answers will vary.
5. Answers will vary.